STUDIES AND TEXTS

51

DOCTOR BERNARD DE GORDON: PROFESSOR AND PRACTITIONER

BY

LUKE E. DEMAITRE

PONTIFICAL INSTITUTE OF MEDIAEVAL STUDIES
TORONTO, 1980

CANADIAN CATALOGUING IN PUBLICATION DATA

Demaitre, Luke E., 1935-
Doctor Bernard de Gordon

(Studies and texts - Pontifical Institute of Mediaeval Studies ; 51 ISSN 0082-5328)

Bibliography: p.
Includes index.

ISBN 0-88844-051-0

1. Bernard de Gordon, fl. ca. 1283-ca. 1308. 2. Medicine, Medieval. I. Title. II. Series: Pontifical Institute of Mediaeval Studies. Studies and texts - Pontifical Institute of Mediaeval Studies ; 51.

R144.B47D45 610'.92'4 C80-094161-6

PONTIFICAL INSTITUTE OF MEDIAEVAL STUDIES
59 Queen's Park Crescent East
Toronto, Ontario, Canada M5S 2C4

PRINTED BY UNIVERSA, WETTEREN, BELGIUM

For father
and in memory of mother

Contents

Preface

Bernard de Gordon, master in the faculty of medicine at Montpellier, is perhaps the most noteworthy representative of the academic quality and professional standards of that *studium* in its golden age (1250-1350). During those years several members and alumni of the university became famous and influential, among them polymaths such as the mystic Ramon Lull and the visionary Arnold of Villanova, one-time law students such as the ruthless Guillaume de Nogaret and the restless Francesco Petrarch, and surgeons at royal or papal courts such as Henri de Mondeville and Guy de Chauliac. But most of these luminaries, with the probable exception of Arnold of Villanova, bequeathed little information about the medical teaching that brought the greatest renown to Montpellier. And all of them, including Arnold, were connected only temporarily or intermittently with the university. The one master, however, who appears to have spent his entire active life at Montpellier and to have devoted all his energy to the teaching and practice of medicine was Bernard de Gordon. His writings cover the full range of the medical curriculum and thus richly document the Montpellier academic tradition. They may also reflect commonly held views more faithfully than do the works of his better known contemporaries, and with perhaps less distortion from egregious exploits and idiosyncrasies, since first-hand external references to Bernard's activities and character are entirely lacking. This means that we depend on internal criteria for the verification of modern references to Bernard as a "leading representative" of orientations as divergent as adherence to the Salernitan legacy and dependence on Arabic authorities, or Anglo-Norman empiricism and dogmatic scholasticism. Upon a closer look at the internal evidence there emerges the portrait of an erudite scholar, a humane teacher endowed with common sense, and a conscientious practitioner. His expositions reveal various features of dialectic and academic medicine, but speculation was not his dominant interest, and his observations and recommendations illuminate many facets of the active life and society around him.

It seems surprising that Bernard de Gordon, who was the subject of a chapter by Émile Littré in the *Histoire littéraire de la France* as well as of paragraphs in most surveys of medieval science and medicine, has thus far not been treated in a full-size monograph. Since his teachings were cited as authoritative at least until the early sixteenth century, not only in Montpellier but also in the medical faculties of Vienna (by Martin Stainpeis) and of Paris (by Jacobus Sylvius), they would appear to deserve closer scrutiny than they have received from students of the university curriculum. Bernard's most noteworthy viewpoints have been utilized for studies on particular medical topics that range from epilepsy and leprosy to dentistry and medicinal quantification, but this still leaves us with a fragmented rather than integrated profile of his thought. Even though the chronology and bibliography of his writings has been unravelled by distinguished historians, from Baron von Töply and Dr. Karl Sudhoff early in this century to Professor Ynez Violé O'Neill more recently, these writings have not yet been brought together for systematic and comprehensive analysis. It is further ironic that no critical editions or modern translations have appeared of Bernard's major works, which were among the prized possessions of such learned bibliophiles as Amplonius Ratinck of Erfurt and Cardinal Nicolaus Cusanus, and which are extant in numerous manuscripts and early printed editions (as may be seen in the appendix to the present study).

The primary purpose of this study, therefore, is to present a comprehensive analysis of Bernard's scholarship and to document that analysis with a wide selection of texts that are available only in manuscripts, in incunabula, or in almost inaccessible editions. Inherent in this purpose are certain pitfalls of which I am apprehensively aware and for which I ask the reader's indulgence. The analysis would have been more thorough if Bernard's views on every subject had been compared systematically with the pre-existing tradition and with the thought of his contemporaries, but I hope that this presentation of Bernard as an author in his own right will be deemed valuable both as a modest contribution to our knowledge of medieval medicine and as the first installment of a more exhaustive investigation. With Bernard we may find comfort in the opening maxim of the Hippocratic *Aphorisms* that inspired so many glossators, "Life is short, the Art long." Specialists in paleography as well as historians of science may find several questions glossed over in this combination of textual research and intellectual synthesis, but my chief concern was not only to discover the individual behind the manuscript folios and the theoretical maxims but also to emphasize the relation between theory and practice.

The footnotes may strike the reader as uneconomically numerous and inordinately long, but they include only the references that are specifically relevant to the discussion and they are intended mainly as editions of the documentary and supporting evidence. In the quoted texts the Latin spelling has been reproduced as found in the manuscripts, and no *sic* has been added when apparent misspellings were common usage in the period, for example *e* for *ae*, *c* for *t* in *-tio* endings, *t* for *ct* in *practica*, *hanelitus* for *anhelitus*, *Ypocras* for *Hippocrates*, and so on.

The present work grew out of my doctoral dissertation, for which my adviser was Professor Pearl Kibre. To her introduction I owe my acquaintance with Bernard de Gordon, to her enthusiasm for learning my interest in medieval thought, to her high standards of scholarship my training in research and writing, and to her warm encouragement my ability to complete this work. In spite of her own busy schedule, Professor Kibre was kind enough to read my manuscript with her usual meticulous care, and I am indeed deeply in her debt for the essence of what may be valuable in this study, while I am entirely responsible for such imperfections and errors as remain. For seeing my work published I am especially grateful to Professor James A. Weisheipl, OP, whose cordial support and expert advice benefited not only the book but also its author.

Without forgetting the many others who have contributed to the growth of this study from inception to publication, I wish to acknowledge my debt to Professor Michael R. McVaugh for his generosity in sharing over these past years much of the important information he has discovered in his research on Montpellier, and for his thoughtful suggestions; to Professor G. Wylie Sypher for his stimulating interest and helpful comments far beyond what is normally expected by a junior faculty member from a chairman – or received by a medievalist from a Renaissance historian –; to Professor Richard Lemay for useful references on astrology; to Mrs. Alice Weaver and her staff in the Rare Book Room of the New York Academy of Medicine for their gracious hospitality in that haven of scholars and for their efficient assistance; to Mr. Jozef Ghyssaert of the Stadsbibliotheek in Bruges, Belgium, and to Msgr. H. Moritz of the Cusanusstift in Bernkastel-Kues, West Germany, for the inspiration I drew from their love of books in a delightfully medieval setting; and also to the librarians in the Colleges at Cambridge and Oxford, the Wissenschaftliche Allgemeinbibliothek in Erfurt, the London British Museum (now Library), the Munich Bayerische Staatsbibliothek, the Paris Bibliothèque Nationale, the Bibliotheca Apostolica Vaticana, the Hessische Landesbibliothek in Wiesbaden, and the Vienna Oesterreichische Nationalbibliothek, for their services in giving me access to indispensable manuscripts.

My thanks to my sons Eugene and Peter, who often restrained their youthful exuberance in order to allow me to concentrate and in deference to my mysterious visitor, the wise doctor from medieval Montpellier. Last but not least, my deepest gratitude goes to my wife: without her loving inspiration this work would not have been undertaken, without her enduring support it would never have been completed, and without her patient cheerfulness it might have been published posthumously.

1

Biography, Educational Background, and *Magisterium*

Bernard de Gordon's biography has to be reconstructed almost entirely from scraps of information scattered throughout the manuscripts of his own works. Aside from these internal clues, documentary evidence on his life is lacking and references by contemporary or later authors are usually to his writings rather than to his person. Nearly all these references are found in medical treatises, but undoubtedly the most famous one occurs in a literary work. Geoffrey Chaucer mentioned Bernard in the Prologue to the *Canterbury Tales*, composed less than a century after Bernard's death.[1] Chaucer lined him up with two Englishmen, Gilbertus Anglicus (early thirteenth century) and John of Gaddesden (ca. 1280–1349). As George Sarton observed, the three "formed a natural group" because they each composed a compendium of practical medicine with a floral title.[2] The *Laurea anglica* of Gilbertus evidently influenced Bernard's *Lilium medicine* and the latter, in its turn, was soon cited in the *Rosa anglica* of John of Gaddesden, who may never have left England. There are further indications that Bernard's work was already well known in the British Isles in the fourteenth century, for example in the number of manuscripts of English provenance.[3] The spread of its fame beyond professional circles

[1]

Wel knew he the olde Esculapius,
And Deiscorides, and eek Rufus,
old Ypocras, Haly, and Galien;
Serapion, Razis, and Avicen;
Averrois, Damascien, and Constantyn;
Bernard, and Gatesden, and Gilbertyn.

Walter Skeat, *The Complete Works of Geoffrey Chaucer*, 2nd ed. (Oxford, 1900), 4: 13, vv. 429-434.

[2] George Sarton, "Lilium medicinae," *Mediaeval Studies in Honor of J. D. M. Ford* (Cambridge, Mass., 1948), p. 241. For a brief comparative analysis of these compendia, see chapter 2 below.

[3] For example, MSS Erfurt Amplonianum (henceforth cited as Ea.) F. 258, before 1310; London British Library Sloane (henceforth BL Sl.) 334, 14th c.; BL Sl. 512, 14th c.; Boston Medical Library 18, 14th c.; and so on.

is poignantly documented by the complaint of a fifteenth-century leech or uneducated practitioner,

> Gordonie of Burgundie that is callid *lilium medisine* ne lanfranc ne suche others newe auctoris, for faute of pecunie, Y may naught geten, thowz Y woot weel where thay been; and pore and symple hath litil helpe.[4]

It is somewhat surprising that the *Lilium medicine* did not become available in an English version in spite of its renown and even though it was translated into Irish twice.[5]

Since his compendium was so renowned in the British Isles, Bernard seems a logical choice for Chaucer's trio of recent medical authorities. It is less clear, however, what exactly, besides the sound of the surname Gordon, earned him the designation of "Scotch Bernard" by a modern translator of the *Canterbury Tales*.[6] This designation appears to have become an established tradition. Dr. Fielding H. Garrison, the author of one of the most widely used textbooks of the history of medicine, called Bernard "inferentially a Scotchman."[7] While this historian cautiously admitted the inferential character of his assumption – and tantalizingly omitted all reference to his source – other authors are far less reserved. Thus another historian of medicine, Benjamin Gordon, perhaps carried away by patriotic pride and clannish bias, asserts that Bernard himself "*claims* to have had Scottish ancestry" – but he also fails to cite a source.[8] When still another author, Donald Campbell, classifies our physician among the "leading representatives of *Anglo-Norman* medicine,"[9] we are

[4] London: MS 124 of the Royal College of Physicians, quoted by Rossell Hope Robbins, "Medical Manuscripts in Middle English," *Speculum* 45 (1970), 409-410.

[5] "... first by a member of the O'Hickey family in the northern part of Munster, then again in the fifteenth century by Cormac MacDuinnsthleibhe, a distinguished scholar and physician of Ulster": George Sarton, *Introduction to the History of Science* vol. 3 (Baltimore, 1947), p. 873 (where no references are given)

[6] *Canterbury Tales*, tr. Nevill Coghill (Baltimore, 1952), p. 37. Muriel Bowden, *A Commentary on the General Prologue to the Canterbury Tales* (New York, 1949), p. 200, stated that Bernard is included among "the important men of [Chaucer's] nation." More recently, John Halverson in his edition, with introduction and notes, of the *Canterbury Tales* (Indianapolis and New York, 1974), p. 18, characterized the trio as "British."

[7] Fielding H. Garrison, *Introduction to the History of Medicine*, 4th ed. (Philadelphia, 1960), p. 164. Most recently, this tradition is still evident in Mario Tabanelli, *Jehan Yperman: Padre della Chirurgia Fiamminga* (Firenze, 1969), p. 36; in Palémon Glorieux, *La Faculté des arts et ses maitres au XIIIe siècle* (Paris, 1971), p. 107; and in Stanley Rubin, *Medieval English Medicine* (London, 1974), p. 196.

[8] *Medieval and Renaissance Medicine* (New York, 1959), p. 283 (italics mine).

[9] *Arabian Medicine and Its Influence on the Middle Ages* (London, 1926), 1: 162 (italics mine).

made fully aware of the potential implications of assumptions with regard to Bernard's actual origins.

Although there is no conclusive evidence to identify the birthplace of Bernard de Gordon, his writings strongly suggest a continental and, more specifically, a meridional origin. One thin clue may be found in his discussion of scrophula, the disease that allegedly could be cured by the touch of the English and French kings. He claims that "kings, and *especially the Most Serene King of the Franks*, were wont to heal by their mere touch."[10] Hence, his own words do indicate a French bias even though Marc Bloch somewhat freely inferred that, "With the more conciliatory spirits," Bernard "did not hesitate to recognize that both dynasties possessed the same miraculous healing power."[11] A clever argument for his possible French roots, derived from the title of his *Lilium medicine*, has been proposed by Charles H. Talbot, who believes that Bernard

> explains that "in a lily there are many flowers; and in each flower there are three petals with six golden stamens. In the same way this book contains seven parts." The three petals probably refer to the Fleur de Lys of France: certainly they could not describe the Thistle of Scotland.[12]

While ingenious, this argument is unfortunately based on a misreading of the *Lilium*'s prologue, "in a lily there are many flowers, and in each flower there are *seven* white petals and *seven* golden stamens."[13] Another misreading may be responsible if Bernard's Scottish parentage was inferred from the surname Gordon, common in Scotland. In the early manuscripts Bernard is usually referred to as *de Gordonio* rather than as Gordo or Gordonius.[14] A more plausible assumption, therefore, is that he was a native of one of the several places in France with the name Gordon or Gourdon. One of these, Gourdon near Montceau-les-Mines (Départe-

[10] "Reges consueverunt curare solo tactu et potissime serenissimus rex francorum, et ideo morbus regius appellatur": *Lilium medicine*, 1.21, in London MS British Library Harley (hereafter cited as BL Harl.) 3698, fol. 12v. Further references to the *Lilium* are to this manuscript, unless otherwise indicated.

[11] *The Royal Touch; Sacred Monarchy and Scrofula in England and France*, tr. J. E. Anderson (London, 1973), p. 84.

[12] *Medicine in Medieval England* (London, 1967), p. 104.

[13] "In lilio enim multi flores sunt, et in quolibet flore sunt septem folia candida et septem grana quasi aurea": *Lilium medicine*, prologue, fol. 1r.

[14] This was already pointed out two centuries ago by Jean Astruc, "Bernardus Gordonius, ou plutôt de Gordonio," in his *Mémoires pour servir à l'histoire de la Faculté de médicine de Montpellier* (Paris, 1767), pp. 176-181. However, see also *Gordonii* in n. 167, and *Gordo* in n. 18 below.

ment Saône-et-Loire), is a small village whose respectable antiquity is suggested by the discovery there of a small treasure from the time of King Sigismund of Burgundy (d. 524).[15] If this village should claim Bernard among its sons, it would find support in the earlier mentioned reference of an English practitioner to "Gordonie of Burgundie." However, historians have been inclined to favor another Gourdon, in Quercy (Département Lot), that lies about two hundred kilometers northwest of Montpellier, the place where Bernard spent most of his life.[16] As Gordonium, in the diocese of Cahors, it was a town of some importance in the thirteenth and fourteenth centuries. By 1261 it was distinguished by the presence of a religious house of the Cistercians and one of the Franciscans.[17] In 1316 King Louis X summoned its burghers to send representatives to a central assembly at Bourges.[18] A further indication of this town's relative importance is found in a Provençal astrological tract of around 1330 that, as we shall see later, may even have some relation to Bernard's own work: in a horoscopic drawing that aligns the zodiac with planets and cities, Gordonium appears together with Bordeaux, Cahors, Toulouse and two other southwestern towns.[19]

Bernard's meridional provenance is most strongly suggested by negative internal evidence. His known writings contain no personal reminiscences but only hearsay information about other regions. In view of his alleged Scottish descent it is especially worth noting that he refers only to England (*Anglia*), together with Arabia as typically "distemperate regions," and that his acknowledged source is the *Regimen sanitatis* of

[15] See Jean Hubert *et al.*, *Europe of the Invasions* (New York, 1969), pp. 218 and 377.

[16] Émile Littré referred to a Gourdon in the Var and one in the Lot: "Bernard de Gordon, Médecin," *Histoire littéraire de la France* (hereafter *HL*), 25 (1869), p. 321. Sarton put Gourdon "in Rouergue (a part of Guyenne; now Département Aveyron)": *Introduction*, 3: 873. For bibliography on this Gourdon, see Ulysse Chevalier, *Répertoire des sources historiques du moyen âge; Topo-bibliographie*, 1 (Paris, 1894), col. 1326; also Pierre Deschamps, *Dictionnaire de géographie ancienne et moderne* (Paris, 1964), p. 574.

[17] Chevalier, *Répertoire*, 1: 1326; Richard W. Emery, *The Friars in Medieval France; A Catalogue of French Mendicant Convents, 1200-1550* (New York, 1962), p. 85.

[18] C. H. Taylor, "Assemblies of French Towns in 1316," *Speculum* 14 (1939), 278. Other towns summoned were Toulouse, Alès, Millau, and Najac. With regard to the last of these towns, there is an intriguing coincidence in the name of one of its representatives, "magister R. Gordo"; see Elizabeth R. Brown, "Assemblies of French Towns in 1316: Some New Texts," *Speculum*, 46 (1971), 286.

[19] Bayonne and Agen are the two other southwestern towns, Paris and Jerusalem the only non-Provençal cities mentioned: Paris MS Bibliothèque Nationale (hereafter BN) lat. [*sic*] 7420A, item 21, described by Paul Meyer, "Traités en vers provençaux sur l'astrologie et la géomancie," *Romania* 26 (1897), 237. See also chapter 2, p. 44, n. 41 below.

Galen of Pergamon (ca. 131–201).[20] The bookish nature of Bernard's knowledge of the North is further illustrated when he mentions the English (*anglici*) together with the Germans (*theotonici*) as having straight blond hair because of the cold climate,[21] and when he describes the Germans as exemplifying the effects of a cold climate on temperament and physiognomy:

> They are good in natural functions but bad in morals, because [they are] warlike, irascible, unrestrained, fierce, always ready to fight, and disobedient to king and law; and because the digestion of the veins is disturbed by the frigidity of the air, they are fat and pale. Nature has also made their horses similar to their own figures, fat and round, as can be seen in the horses of the Germans.[22]

Such stereotyped characterization, evidently derived from some encyclopedia or from a commentary on Ptolemy's *Quadripartitum* rather than from personal observation, prevails throughout Bernard's treatment of the four climates and of foreign nations. Hence we should not take too literally his statement that visual observation of Germans proves that northern people are taller, stronger and braver,[23] or his claim about their antipodes in the far south, "the Pygmies, who are less than one cubit and one half palm in height, as I have seen two."[24]

Even for the nearby southern countries Bernard's information was clearly indirect. To Spain he alluded only vaguely, once as a spawning ground for philosophers because it shares with Greece a place in the ideal

[20] "Dicit Galenus in de regimine sanitatis, particula prima, et intendit dicere quod in regionibus distemperatis non invenitur corpus optimum et exemplificat in anglia et arabia dicens quod in regionibus multum distemperatis non possunt inveniri corpora temperata": Bernard de Gordon, *De conservacione vite humane*, 4: *Regimen sanitatis*, 11, in Vatican MS Palatine Latin 1174 (hereafter Pal.), fol. 66v (references to the *De conservacione* hereafter are to this MS, unless otherwise indicated).

[21] "Indi et etiopes habent [capellos] nigros et crispos, et anglici et teutonici flavos et planos": *Lilium medicine*, 2.1, fol. 21r.

[22] People in northern regions "sunt autem boni in naturalibus, mali autem in moribus, quoniam bellicosi, irascibiles, indomabiles, furiosi, preparatissimi ad arma, inobedientes regi et legi; et quia digestio venarum propter frigiditatem aeris corrumpitur, ideo pingues, albi. Et cum hoc natura fecit equos similes figuris eorum, pingues et rotundos, ut patet in equis alemanorum": *Liber pronosticorum*, 2.7, in Vatican MS Pal. 1174, fol. 84r. (references hereafter to this MS).

[23] "Ad sensum enim videmus quod [in regionibus distemperatis in frigiditate homines] sunt maiores in corpore et forciores et magis audaces sicut patet in theotonicis": *De conservacione*, 4: *Regimen sanitatis*, 11, fol. 66v.

[24] "Ultra autem regionem, ubi colligitur piper, accidunt multa monstra et peccata, sicut sunt Pygmei, qui non sunt ad longitudinem unius cubiti palma cum dimidia, sicut ego vidi duos": *Liber pronosticorum*, fol. 84r.

"fourth climate,"[25] and once as the southwestern region where one finds "choleric, swift and nimble" people and an Arab nation "called Geneth in the vernacular."[26] About Italy he reported a few unflattering things, but in each instance he admitted that his knowledge was based on hearsay or even gossip. Thus, he stated that many devilish recipes for lethal poisons "*seem* to be known among many Italians, so much so that they even put poison in the consecrated host, and they kill without fail *as the common rumor goes*."[27] On another unpleasant subject, he related that a mole of the uterus or a mooncalf was called *frater lumbardorum* because

> *it is said* that this occurs quite often in women of Lombardy, either because they work harder as they are more submissive, or because of inappropriate nutrition as in Apulia, for *it is commonly said* that Italian women live quite poorly, on fruits and herbs, but that they dress well.[28]

Bernard's secondhand and often disparaging comments about non-Provençal regions and people become a more convincing argument for his meridional origin when they are contrasted with the directly personal and favorable nature of his references to Provence, even though in itself this aspect might be explained by the fact that all his works were written in Montpellier. Avignon was the only place which he called "nearby (*prope nos*),"[29] and of the six metropolises he listed – as suffering from air pollution – two were distant (Rome and Paris) but four were meridional:

[25] "Manifeste videmus quod philosophi fuerunt plurimi in quarto climate quam in alio sicut fuerunt a parte orientis in grecia et a parte occidentis in parte yspanie in corduba": *De conservacione*, 4: *Regimen sanitatis*, 11, fol. 66v.

[26] "In parte tamen occidentali, que declinat versus meridiem, versus africam, homines sunt ut plurimum cholerici, mobiles, veloces in motu, cum faciebus oblongis ... sicut patet ad sensum in gente yspanie, que approximantur [*sic*] ex uno latere versus meridiem mari mediterraneo et ex altero oceano, ut patet in gente arabum dicta vulgariter geneth": *Liber pronosticorum*, fol. 83v.

[27] "Et omnes isti modi et plures alii videntur esse noti apud plures Italicos in tantum eciam quod venenum ponunt in hostia consecrata et interficiunt irremediabiliter sicut fama communis currit et circa antiquos non consueverunt currere et currunt ista venena potissime circa prelatos domino permittente propter culpam nostram": *De tyriaca*, ed. J. Pagel, in *Pharmaceutische Post*, 27 (1894), 258.

[28] "Frater lumbardorum" or "mola matricis frequentius ut dicitur generatur in mulieribus lumbardie, aut quia magis laborant quia magis sunt subiecte aut propter corrupcionem nutrimenti, sicut in Apulia, quia dicitur communiter quod Italice pessime vivunt scilicet ex fructibus et herbis, sed bene induunt": *Lilium medicine*, 7.18, in the Venice incunabulum (1498), fol. 93v. The MS version, in Paris BN lat. 6964 (early 14th c.), fol. 78v, has "omnes Ytalici pessime vivunt," while London BL Harl. 3698, fol. 96v has "sicut in Apulia ut aliqui voluerunt," and omits the rest, which may suggest a sensitive Italian scribe.

[29] *Lilium medicine*, 1.14, fol. 8r.

Lyon, Arles, Toulouse and Narbonne.[30] The *Provincia Narbonensis* or Provence, "this province" as he called it, was the only area with which he proved thoroughly familiar and to whose wines, incidentally, he gave special attention.[31] Moreover, when he occasionally gave a synonym for some particular wine, vegetable, concoction, or affliction, the Provençal vernacular (*vulgare*) or dialect (*ydioma*) was the only language, other than Latin, of which he demonstrated any knowledge.[32] When considered individually, such references may appear inconsequential; but when combined, they seem to provide persuasive evidence for the assumption that Bernard hailed from Provence. However, this assumption should not lead one to confuse him with another physician, Bernardus Provincialis of Arles, who lived more than a century earlier.[33]

Even if he was born on a distant shore, Bernard de Gordon had become a truly "naturalized" denizen of Provence and citizen of Montpellier. In fact, we even discover a tinge of chauvinism when he claims that "the very beautiful Montpellier region" is situated in a temperate area,[34] for he believes that such areas are distinguished by their "temperate, just,

[30] Bad airs occurs "potissime in magnis civitatibus ubi aer est caliginosus et humidus cum sunt supra magna flumina constitute, sicut est Roma domina provinciarum, Lugdunum, Civitas Arelatensis, Parisius, Tholosa, Narbonensis et ita de aliis": ibid., 5.8, fol. 66r.

[31] "[Vinum] in ista provincia quod fieret ... de uvis candidis que nominantur uve columbares," *De marasmode*, in Vatican MS Pal. 1234, fol. 139v. "Vina que crescunt in Monte Pessulano et ut plurimum in tota provincia Narbonensi": *De conservacione*, 4: *Regimen sanitatis*, 12, fol. 67v. "Galenus nec Ypocras numquam viderunt vinum album fortis virtutis et tamen hodie invenitur quasi sicut aqua ardens cum virtute quasi intolerabili et ista inveniuntur in multis locis in provincia Narbonensi": ibid., 5, fol. 58v. The term Provence was often applied to the entire Midi, even though in more precise usage it refers to the area east of the Rhône while the area west of the Rhône (including Montpellier) is called Languedoc, a term not used by Bernard de Gordon.

[32] "Vinum anni citrinum in colore, dictum vulgari provincialium castaynch, mediocre in substancia, odoriferum": *Liber pronosticorum*, fol. 82r. Against hernia, "Rx fabe inverse que vulgari provincialium vocatur benadus": *Lilium medicine*, 7.7., fol. 91v. Against consumption "utatur confectione que dicitur tri. [?] sed in ydiomate provincialium menudeli": ibid., 4.5, fol. 52v. Certain pustules "possunt ita vocari in vulgari provincialium broshadaur [?]": ibid., 1.27, fol. 17r.

[33] "Le Bernard 'provençal', qu'on trouve cité, est vraisemblablement le même que Bernard de Gordon": Littré, *HL* 25: 321. See also the inference of Lynn Thorndike, *History of Magic and Experimental Science* 1 (New York, 1923), 740. For Bernardus Provincialis, native of Arles and student of Salerno, who lived around the middle of the twelfth century, see Ernest Wickersheimer, *Dictionnaire biographique des médecins en France au moyen âge* (Paris, 1936), 1: 78 and 2: *addenda*.

[34] Temperaments are different "in regione calida ut in Aethyopia et aliter in frigida ut in Teutonica et aliter in regione vergente ad temperamentum inter ista sicut est preclara regio Montispessulani": *De gradibus*, ed. J. Pagel, *Pharmaceutische Post*, 27 (1895), 133.

handsome and very wise people."[35] One particular group of people in that region, however, may have been less healthy and suffering from an endemic blood disease that fascinated Bernard intensely. He related,

> I have seen among certain people (*apud quasdam gentes*) blood that had an off-white, ashen color like tallow, and I have seen this in every age group of that territory, in both sexes, and in most every season; and I have found this to be a great marvel.[36]

The late Dr. Karl Sudhoff, one of the most distinguished historians of medieval medicine, interpreted this description as referring to a "northern land" and as possibly confirming Bernard's Scottish descent.[37] However, Bernard himself a few chapters later identified the place of his hematological discovery as Montpellier:

> This blood appears in general in Montpellier because those people are usually indulging in idleness and too much food and drink, so that the digestion is impaired and the chyle is not transformed into the true form of the humours; such color, therefore, naturally indicates impairment. I have sedulously investigated and inquired about that color from the phlebotomists, and I have found such blood generally among this group (*apud istam gentem*). Therefore, when the color appears ashen or off-white, or like the color of tallow or of mushrooms, the diet and regimen should be changed.[38]

This passage from the treatise *De flebotomia* (1308) has been quoted at some length because it is intriguing not only to the hematologist but also to the historian who endeavors to reconstruct Bernard's background.

[35] "Regiones temperate cognoscuntur ex effectibus temperatis. Unde illa regio universaliter est temperata, in qua inveniuntur homines temperati, iusti, electe forme, magne prudencie...,": *Liber pronosticorum*, 2.7, fol. 84r.

[36] "Vidi enim apud quasdam gentes sanguinem habentem colorem subalbidum subcinericium sicut si esset sepum, et vidi hoc in omni etate illius territorii et in omni sexu et in omni tempore ut plurimum, quod reputavi prodigium grande": *De conservacione*, 1: *De flebotomia*, prologue, fol. 10r. See also p. 61 below.

[37] "Leider sagt er nicht, in welchem nordischen Lande er diese Beobachtung zahlreicher Fälle leukämischen Blutes gemacht haben will. Er selbst wird ja, wohl nicht ganz mit Unrecht, für schottischer Abkunft gehalten": Karl Sudhoff, "Zur Schriftstellerei Bernhards von Gordon und deren zeitlicher Folge," *Archiv für Geschichte der Medizin*, 10 (1917), 175, n. 3.

[38] "Talis sanguis apparet ut plurimum in monte pessulano quia ut plurimum sunt homines qui sunt in ocio et sunt in multitudine cibi et potus et ideo corrumpitur digestiva et chilus non transmutatur ad veram formam humorum; talis ergo color naturaliter corrupcionem significat, et investigacione huius coloris fui sedulus inquisitor apud flebotomatores et ut plurimum inveni talem sanguinem apud istam gentem; cum ergo apparet color subcinericius aut subalbidus aut sicut color sepi aut fungi, mutetur dieta et regimen": *De conservacione*, 1: *De flebotomia*, fol. 20v. A somewhat free French translation is given by Littré, *HL* 25: 330.

Contrary to what Dr. Sudhoff asserted, the text contains no allusion to the North and the observation is related directly to the region in which Bernard was working and writing. It is conceivable, furthermore, that the references to a specific group, *istam gentem*, of a separate district, *illius territorii*, may point towards the inhabitants of the Jewish quarter in Montpellier or a nearby territory. Bernard had used the phrase "usually in idleness" in speaking of the Jews five years before he recorded his findings in the work on phlebotomy. In his *Lilium medicine* he alleged that "the Jews generally suffer from bleeding hemorrhoids" for the following three reasons:

> they are usually in idleness and therefore they build up excessive melancholic humours; in the second place, they are usually in fear and anxiety, and thereby their melancholic blood is increased, according to Hippocrates' maxim that "fright and despondency, if lasting for a long time, is a melancholic affection" [*Aphorisms* 6.23]; the third reason is on account of divine punishment, according to the saying, "He struck them in the posterior and gave them perpetual opprobrium" [Psalm 77.66].[39]

Bernard's uncritical adherence to gross popular myths about Jewish shameful ailments and blood loss – myths that were underlying factors in the superstition that the Jews wanted Christian blood[40] – makes an even more poignant impression because it is combined with his perceptive appreciation of their constant fear. Furthermore, he had provided evidence that their fear was not unfounded, when in a previous chapter he remarked very casually that the food-attracting powers of the stomach are demonstrated "in the Jews when they are hung by their feet."[41]

Statements like these rule out another possible hypothesis about Bernard's origin, namely that he might have been Jewish. They invalidate that hypothesis more unequivocally than the reasons offered by George

[39] "Advertendum quod Iudei ut plurimum paciuntur fluxum emorroydarum propter tria: et quia communiter sunt in ocio et ideo aggregantur sibi fluiditates melancolie; secundo quia communiter sunt in timore et anxietate et ideo multiplicatur sanguis melancolicus iuxta illud Ypocratis, Timor et pusillanimitas si multum tempus habuerint melancolicum faciunt [*Aphorisms* 6.23]; tercio quod hoc est ex ulcione divina iuxta, Et percussit eos in posteriore dorsi, opprobrium sempiternum dedit illis [Psalm 77.66]": *Lilium medicine*, 5.21, fol. 77r. The reference in Psalm 77 is actually to the hemorrhoids inflicted upon the Philistines for detaining the Ark.

[40] See Joshua Trachtenberg, *The Devil and the Jews; The Medieval Conception of the Jew and Its Relation to Modern Antisemitism* (Cleveland and New York, 1961), pp. 50, 147-148, and references.

[41] "Et sicut vidimus in Iudeis cum suspenduntur per pedes, nichilominus tamen cibus attrahitur ad stomachum": *Lilium medicine*, 5.2, fol. 61r.

Sarton, who argued that medieval Jews did not use the name Gordon and especially that "Jews were at that time unwelcome in Montpellier."[42] With regard to the latter argument it should be pointed out that the situation of the Jews in Montpellier was more complex. On the one hand, the precariousness of their life, which we find reflected in the tenor and tone of Bernard's remarks, culminated in their expulsion from Montpellier and Roussillon by Jayme II of Mallorca in 1307.[43] The anti-Judaic sentiment was apparently fanned by some members of the university, including Arnold of Villanova (ca. 1240–1311), who also complained to Pope Boniface VIII (r. 1294–1303) about the fact that the French prelates were attended by Jewish physicians.[44] At the same time, however, there was frequent cooperation between Christian and Jewish scholars in Montpellier. The cynosure of this cooperation was the scholarship of Jacob ben Makhir, who is better known as Profatius Judeus (ca. 1236–ca. 1305).[45] Bernard de Gordon himself requested Profatius to draw up and explain an armillary or concentric map of the celestial sphere.[46] The collaboration between Profatius and masters of the university was most intense in the field of astronomy, but it also yielded an occasional bonus to medicine proper. In 1299, for example, Profatius translated Avenzoar's *Taisir* or *Regimen sanitatis* with the assistance of the surgeon Bernardus Honofredi and at the request of the physician Petrus de Capitestagno.[47] While this Judeo-Christian cooperation climaxed in the closing years of the thirteenth century, it had flourished for

[42] Sarton, "Lilium medicinae," p. 242, n. 3.

[43] By this expulsion Jayme II followed the example set by Philip IV the Fair in 1306.

[44] Jacques Pinès, "Des médecins juifs au service de la Papauté du XIIe au XVIIe siècle," *Revue d'histoire de la médecine hébraïque*, 69 (1965), 126, where no specific reference is given. See also Joaquin Carreras y Artau, "Arnaldo de Vilanova, apologista antijudaico," *Sefarad*, 6 (1947), 49-61.

[45] Some noteworthy products of this cooperation are mentioned in my article, "Theory and Practice in Medical Education at the University of Montpellier in the Thirteenth and Fourteenth Centuries," *Journal of the History of Medicine and Allied Sciences*, 30 (1975), 108-109. Profatius' collaborations with Christian scholars do not warrant the assertion that he became chancellor of the University, as made by Paul Delmas and Charles Guérin-Valmale, "L'Hippocratisme montpelliérain; ses origines; Rôle prépondérant des Fils de l'Islam et des Enfants d'Israel." *Aesculape* 13 (1923), 94.

[46] Paris MS BN lat. 7295A (14th c.), fol. 40v, *desinit*: "et sic completum est opus nostrum ad peticionem venerabilis magistri Bernardi de Gordonio doctoris excellentissimi in arte medicine in Monte Pessulano. Explicit de armillis Profacii."

[47] See Wickersheimer, *Dictionnaire*, 1: 76 and 2: 620. For manuscripts see Lynn Thorndike and Pearl Kibre, *A Catalogue of Incipits of Mediaeval Scientific Writings in Latin*, rev. and augmented ed. (Cambridge, Mass., 1963; hereafter cited as TK), col. 187; other 13th-century translations of the *Taisir*: TK 406 and 751.

several decades. Thus, thirty six years before the 1299 team project just mentioned, Profatius and master Johannes de Brescia had collaborated on a Latin translation of al-Zarqalī's *Saphea* or astronomical table.[48] However, that was in 1263, when Bernard was probably learning just his first Latin words.

The question of Bernard's education, much more than that of his national origin, has potential implications for his exact place in the academic context of the late thirteenth century. Since here, too, we depend entirely on his own writings, all of which were related to medicine, we can only speculate about the setting as well as the nature or content of his premedical education. The Montpellier region, or at least the Midi, seems plausible as the place if we consistently use, as an argument *ex silentio*, the absence of personal reminiscences about other areas in his geographical references. Bernard's silence is less helpful, however, where he fails to inform us about any of the subjects which he studied before entering medical school. It is safe to assume that he received the training in the liberal arts that was emphasized as a prerequisite for the study of medicine. The importance of the arts for medicine was emphasized not only in the general tradition, with its well known epitome by Isidore of Seville (ca. 560–636),[49] but also in the thirteenth- and fourteenth-century statutes of the universities, including Montpellier.[50] With regard to specifics, however, the problem of Bernard's silence is compounded by the close institutional alliance between medicine and the arts and by the lack of details in the curricula of both the faculty of arts and the faculty of medicine at Montpellier.[51] Hence, we depend on Bernard's general references to education and on his citations of nonmedical sources for our attempt to draw up a profile of his erudition.

Bernard believed that a boy should be "put to letters and should learn to read, to write and to count" at the age of six, although the child "should not be forced to stay in class too long."[52] While some southern French

[48] Sarton, *Introduction*, 2: 851. This translation was ed. by J. Millas-Vallicrosa (Barcelona, 1933). See also Thorndike, *Magic*, 3: 201; and Emmanuel Poulle, "Un instrument astronomique dans l'occident latin, la 'saphea'," *Studii medioevali*, 10 (1970), 1: 491-510.

[49] *Etymologiae*, 4.13.1-5, tr. William Sharpe, *The Medical Writings of Saint Isidore* (Philadelphia, 1964), p. 64.

[50] *Cartulaire de l'Université de Montpellier*, 1 (Montpellier, 1890), Documents 2, 4 and 5 (references hereafter are to this volume).

[51] Ibid.,; see also my article, "Theory and Practice," pp. 106-107.

[52] "Postquam pervenit ad vj annos ponatur ad literas et discat legere et scribere et

universities, such as Toulouse, apparently admitted eight or nine-year old students,[53] Bernard's educational program put the crucial stage at the age of twelve. At that age a boy "must choose the way of life which he intends to pursue or towards which he is inclined," namely either a military career, or one of the crafts or trades, or the pursuit of learning. For the first two walks of life, the preparation should concentrate on the development of stamina through vigorous exercise. If the boy prefers military service, *miliciam*, "so that he may be able to defend the commonwealth manfully," he must not only acquire strength and fortitude but also "become accustomed to seeing people who have died in war"; if his choice is a craft or a trade, *ars mechanica* or *ars negociatoria*, "without which we cannot have the necessities of life," the aspiring fisherman should get used to cold water and the future merchant should practice walking. For the third career, however, *ad philosophiam*, it is better to prepare by "gentle and moderate exercise"; the pursuit of learning, *via speculativa*, "when one wishes to train in the sciences and philosophy," requires a mild temper and tranquillity.[54]

It is of interest to note that, in this one paragraph, Bernard's characterization of the third walk of life ranges from the broad *via speculativa* to the more academic term *philosophia* and to the more precise "training in the sciences and philosophy." Moreover, in a parallel passage in an earlier work he assigned vocational differentiation to the age between seven and fourteen, and he seemed to narrow down the third career even further to "letters" (*literas*).[55] This range adequately reflects the bewildering complexity and diversity of thirteenth-century schemes of learning.[56]

---

numerare et non cogatur diu stare in scolis": *De conservacione*, 4: *Regimen sanitatis*, 5, fol. 57v.

[53] See Louis J. Paetow, *The Arts Course at Medieval Universities* (Urbana, Ill., 1910), p. 55, n. 16.

[54] "Cum autem pervenerit ad etatem annorum xij, tunc debet eligere viam quam intendit vel ad quam magis inclinatur iuxta concursum particularium; tunc igitur aut declinat ad miliciam aut ad artem mechanicam aut ad philosophiam. Si autem appetit miliciam, ut possit rem publicam viriliter defensare, tunc debet adquirere robur et fortitudinem ... assuescat igitur videre homines mortuos in bello.... Secunda via est ars mechanica et ars negociatoria sine quibus vite necessaria haberi non possunt.... Tercia autem via est via speculativa cum aliquis vult se exercitare in scienciis et philosophia, et iste indiget exercicio suavi temperato ... et [isti] inter omnes homines magis debent esse mites et in quiete": *De conservacione*, 4: *Regimen sanitatis*, fol. 58r.

[55] In the third age which starts at 7 and "durat fere usque ad xxiiij [*sic*, read xiiij] annos," boys "debent poni ad officium quod intendunt vel ad quod magis inclinant, ut ad literas vel ad artem mecanicam [et miliciam] et consimilia, sicut melius dicemus in libro de regimine sanitatis": *Liber pronosticorum*, 2.6, fol. 83r.

[56] See the articles by Philippe Delhaye, Pearl Kibre, and James A. Weisheipl in *Arts*

Bernard's *via speculativa* might include the religious life, while *philosophia* could embrace not only the three philosophies – rational (the trivium), natural (the quadrivium, physics, and metaphysics), and moral – but also theology. His phrase "the sciences and philosophy," on the other hand, is possibly a specific term denoting the liberal arts of, respectively, the quadrivium (geometry, arithmetic, astronomy and music) and the trivium (grammar, rhetoric and logic). His narrower reference to *literas*, finally, may be broadened on the basis of an analogous instance in the Montpellier documents. In the 1242 statutes granted to the faculty of arts by the bishop of Maguelone only grammar and logic were mentioned[57] but, as the editor of the documents has pointed out, this pair also comprised rhetoric and the quadrivial arts.[58] In the same vein we may observe that, even though in this context Bernard used the term "art" only for a craft (*ars mechanica*) and for a trade (*ars negociatoria*), he can hardly have been unaware of the notion of the *artes liberales*.

The preceding considerations about Bernard's ideas on the third walk of life may serve as the framework for a sampling of his premedical erudition. We should, of course, proceed with the reservation that, in the wide spectrum of quotations, his firsthand knowledge of the sources can rarely be isolated from the instances where he borrowed from commonly available lore. If we begin with the liberal arts, we find the trivium represented when Bernard cites such works as the *Fasti* and the *Remedia amoris* of Ovid,[59] the *De duodecim cesaribus* of Suetonius,[60] and the *De consolatione philosophie* of Boëthius.[61] These works were often found in the medieval grammar curriculum.[62] Bernard's literary background may

---

libéraux et philosophie au moyen âge; Actes du IVe Congrès international de philosophie médiévale (Montreal, 1969), especially pp. 166, 171-172, 180-181, 191 and 209.

[57] *Cartulaire*, document 6.

[58] Ibid., introduction, p. 27.

[59] "Dicebat Ovidius in De fastis, Quid creat omne genus volucrum nisi blanda voluptas?": *Lilium medicine*, 7.2, fol. 90r. Bernard also quotes from Ovid's *Remedia amoris* and mentions his *De arte amandi*.

[60] "Legitur enim in libro De xii cesaribus quod augustus cesar filiam et nepotes fecit sic instrui": *De conservacione*, 4: *Regimen sanitatis*, 6, fol. 60r. The same work is probably meant when Bernard refers to Caesar's death as "legitur in historiis imperatorum": *De tyriaca*, ed. Pagel, p. 259.

[61] "Agamus ergo iuxta consilium boecii in fine De consolacione, Adversamini vicia...,": *De conservacione*, 4: *Regimen sanitatis*, prologue, fol. 54v. Another reference to Boethius is perhaps to the pseudo-Boethian *De disciplina scholarium*: "[dicit] Boecius vinum modice sumptum intellectui videtur acumen inferre": *De conservacione*, 4: *Regimen sanitatis*, 6, fol. 59r.

[62] See Ernst Curtius, *European Literature and the Latin Middle Ages*, tr. W. Trask (New York, 1963), p. 50.

be further illustrated in his extensive quotations from poetry as well as from the Bible. As to the former, the quality of his poetic selections varies widely. Thus, a philosophical aphorism from Aristotle's *Physica* is sandwiched between such common Latin maxims as "*Quod nova testa capit inveterata sapit*" and "*Qui bonus est in studio pravus est in foro.*"[63] There are also more folksy proverbs, one of which reminds us of the nursery rhyme "a frog he would a wooing go."[64] Or a verse with a more classical flavor is attributed to an unnamed poet,[65] while occasionally several lines are borrowed from a *versificator* – a label applied not only to authors of versified grammar textbooks but also to one of their most erudite twelfth-century masters, namely Bernardus Silvestris.[66] In most cases, however, the *versificator* would be cited for a medical mnemonic doggerel, of which we will see some examples in a later chapter. It would seem, too, that Bernard culled primarily the poetic lines from the Bible. He drew especially on the Wisdom Books for his inspiration, and he called the author of Ecclesiastes simply *Sapiens*.[67]

With regard to the second art of the trivium, rhetoric, Bernard reminded us that "wisdom without eloquence is of little use, as written in the prologue to the *Rhetorica*."[68] This reference is to the oldest Latin

[63] "[Pueri] fortissime memorantur, iuxta illud, Quod nova testa capit inveterata sapit; religiosi bene memorantur ... iuxta illud, Anima sedendo et quiescendo fit prudens [Aristotle, *Physica*, 7.4, 247b-248a], et iuxta illud, Qui bonus est in studio pravus est in foro": *Lilium medicine*, 2.12, fol. 27v. For the first of these maxims, compare Horace, *Epistles*, 1.2.69; its medieval currency is evident from the number of references in Hans Walther, *Proverbia sententiaeque Latinitatis Medii Aevi* (Göttingen, 1963-1967), item 25948. The third maxim is cited as "Qui bonus est in studio, pravus est in foro et in generatione vel venereo actu," in the *reportatio* by Conradus de Austria of Albertus Magnus' 1258 *Quaestiones super De animalibus*, 18 q.4, ed. by Ephrem Filthaut in *Opera Omnia Alberti*, 12 (Münster, 1955), p. 299. This maxim is not listed by Walther, but compare, "Bonus scholasticus, malus politicus": *Proverbia*, item 2138.

[64] "Si quis amat ranam putat esse dianam": *Lilium medicine*, 2.19, fol. 31r. Numerous references in Walther, *Proverbia*, item 28967.

[65] For example, "dicebat poeta, Senex fidelis prima credendi via": *De conservacione*, 4: *Regimen sanitatis*, fol. 53v.

[66] "Auditus est porta intellectus, et propter hoc versificator dicit, Quicquid Roma legis, quicquid studuistis Athenis, quicquid ad elencos arguto disputat ore Gallus, et in medica iactitat arte Ligur, quicquid Aristoteles divino pectore sensit, cessit ab auditu, docilis prudensque perisset litera si surdis auribus esset homo": *Lilium medicine*, 3.9, fol. 42r. Compare Bernardus Silvestris, *Cosmographia*, poem 14, ed. Peter Dronke (Leiden, 1978), p. 152.

[67] For example, "dicit Sapiens, Qui addit scienciam addit et laborem [Eccles. 1.18]": *De conservacione*, 4: *Regimen sanitatis*, fol. 54r.

[68] "Scribitur in prologo rhetorice, Sapiencia sine eloquencia parum prodest": ibid., fol. 58v.

textbook for rhetoric, the *Rhetorica ad Herennium* (ca. 85 BC), which was usually attributed to Cicero and called *Rhetorica nova*, rather than to Cicero's *De inventione* which, ironically, was called *Rhetorica vetus* during the Middle Ages.[69] Further evidence of Bernard's proficiency in the art that was also taught in medieval schools as *ars dictaminis* may be found in his style, especially in the introductions to several of his treatises. A few lines from the preface to his *Lilium medicine* may serve as an example. He opens with a reference to that most eloquent and venerable sage, Socrates,[70] who, "questioned by someone as to how he could speak best, answered, 'If you say only what you know well enough.'" Then Bernard announces that his compendium of practical medicine is addressed to the humble rather than to the snobs who

> disdain simple writings, nay blush to learn from what has been said once. And yet Seneca states that, what cannot be said enough cannot be said too much. And Horace [says], "Things repeated ten times will please." Hence, since man's memory is unstable, I am not ashamed to repeat for the humble what is commonly taught about [medical] practice.[71]

It is worth noting, in this excerpt, the theme of repetition and the emphasis on memory, since *memoria* was the fourth of the five divisions of rhetoric as an *ars*.[72] In fact, Bernard seems to have become something of an authority on the subject, since his discussion of the impairment and care of memory in the *Lilium medicine* (2.12 and 13) was cited together with Aristotle's *De memoria et reminiscentia* in the *Sermones medicinales* of the Florentine Niccolò Falcucci (d. 1411),[73] and in the *Practica* or

[69] See Curtius, *European Literature*, pp. 66 and 153.

[70] Plutarch's *Moralia* may be the source of this Socratic dictum, as well as of Bernard's line about people who "vivunt ut comedant" (*Lilium medicine*, 5.8, fol. 66v), which is also attributed to Socrates by Plutarch, *Moralia*, 4.

[71] "Interrogatus a quodam Socrates quomodo posset optime dicere, respondit, Si nichil dixeris nisi quod optime sciveris.... Et quia humilibus scribo, ex hoc superbi abiecti sunt, ... scripta communia contempnunt, immo erubescunt dicere aliquid ex quo semel dictum est. Et tamen dicit Seneca quod numquam nimis dicitur quod satis non dicitur. Et Oracius, Decies repetita placebunt. Quia igitur memoria hominis labilis est, quod communiter de practica dicitur repetere humilibus non erubesco": *Lilium medicine*, fol. 1r.

[72] See Curtius, *European Literature*, p. 68. The classical five parts of rhetoric (invention or analysis, disposition or organization, composition or style, memory, and delivery) were cited by such standard authors as Boethius, Isidore and Alain de Lille: see *Readings in Medieval Rhetoric*, ed. Joseph M. Miller, et al. (Bloomington, Ind., 1973), pp. 26, 71, 75, 81, 128 and 224.

[73] Nicolaus Falcutius, *Sermones medicinales; Sermo tertius de membris capitis* (Venice, 1490), fols. 12 and 68 (at the New York Academy of Medicine).

Philonium of the Portuguese Valescus de Tharanta (1382–1417), a member of the University of Montpellier.[74] Furthermore, the Aristotelian principles, that memory was aided most by a cognitive progress from the universal to the particular and by an orderly sequence of subjects,[75] were faithfully observed in the entire *Lilium medicine* itself. As indicated in Aristotle's views on memory, there was a close connection between rhetoric and the third art of the trivium, logic or dialectic. When we examine Bernard's writings in the following chapters, we will find his training in this art amply illustrated in the logical organization of the subject matter and in the dialectical arrangement of *questiones*, contradictory *opiniones*, and conclusions. He also refers frequently to Aristotle's *Topica*,[76] a work that deals with syllogistic arguments from probable premisses and that, as part of the "New Logic," was lectured on in university courses.[77]

Of the classical authors who constituted the gallery of the medieval curriculum in the trivium Bernard mentioned by name the following: Pythagoras, Sallust, Horace, Ovid, Seneca, and Boethius. While Ovid and Seneca figure most prominently (nine citations each), one cannot fail to notice the absence of such curriculum classics as Cicero or Virgil. Of the early Christian authors, who were usually included under the disciplines of the trivium, Bernard names Origen, John Chrysostom, Jerome, Augustine, Prosper of Aquitaine, Fulgentius, and Pope Gregory the Great. Here Saint Augustine (seven citations) and Saint Jerome (six citations) outnumber the others, while such authors as Cassiodorus and Isidore are conspicuously absent.[78]

Bernard's selection of more recent nonscientific authors may perhaps be considered peculiar. He cites the bawdy twelfth-century comedy *Pamphilus*, warning that it should not be read by children under twelve.[79]

[74] Valescus de Tharanta, *Practica que Philonium dicitur* (ed. by Nicolaus Wolff, no place given, 1500), fol. 23v (at the New York Academy of Medicine).

[75] The Aristotelian ideas were summarized around 1300 by Johannes de Sancto Amando in the prologue to his *Revocativum memorie*, ed. by Otto Paderstein (Berlin dissertation, 1892), p. 3.

[76] For example, "iuxta illud Aristotelis in topicis, Nemo eligit iuvenes duces eo quod non constat prudenciores": *De conservacione*, 4: *Regimen sanitatis*, prologue, fol. 53v.

[77] Thus, at the University of Toulouse in 1309 according to Paetow, *The Arts Course*, pp. 96-97.

[78] Isidore's *Etymologies* may be the source of some of Bernard's rather strange etymologies of such terms as theriac, gonorrhea, and *amor hereos* or lover's malady.

[79] "Non debet audire pamphilum nec libellum de arte amandi Ovidii": *De conservacione*, 4: *Regimen sanitatis*, 5, fol. 57v. *Pamphilus de amore*, by an anonymous

The only "literary" moderns, whom he mentions by name, are Hugh of Saint Victor and Bernard of Clairvaux. It would seem, however, that Bernard de Gordon, while rather careful in identifying the classics and the Fathers whom he quotes, is more liberal in citing modern authors. Perhaps he assumes that the reader knows these moderns well enough to recognize their quotations. Thus, he seems to use the *De proprietatibus rerum* (ca. 1230–1240)[80] of Bartholomeus Anglicus on several occasions,[81] although he mentions it only once – and even then without naming the author.[82] This apparent carelessness reminds us of the basic problem in tracing Bernard's literary background, that is, the difficulty of determining whether he draws directly from an original source or from a florilegium or an encyclopedia. The problem is compounded further by the fact that the encyclopedists themselves drew upon florilegia as well as upon original sources.[83]

In spite of the above reservation, there seems to be sufficient evidence to consider Bernard accomplished in the arts of the trivium. The evidence is more ambiguous, on the other hand, with regard to his schooling in the arts of the quadrivium whose importance for the trained physician was stressed by curriculum authors.[84] The teaching of these mathematical arts in most thirteenth-century schools is poorly documented, although this does not mean that it was neglected.[85] In the case of the Montpellier faculty of arts, the statutes do not mention the quadrivium and the extant works of its members relate exclusively to the trivium. Hence Alexandre Germain wondered whether any quadrivial arts were taught at Montpellier, and he found further reason for skepticism in his interpretation of

---

12th-century author, was based on an elegy in Ovid's *Amores*, 1.8. It was listed among the textbooks of rhetoric in Eberhard the German's *Laborintus* in the early 13th century: Curtius, *European Literature*, pp. 50 and 386 n. 17. The *Pamphilus* is probably the origin of the word "pamphlet."

[80] For the date and character of the *De proprietatibus rerum*, see Thorndike, *Magic*, 2: 401-435 and Sarton, *Introduction*, 2: 586.

[81] For example, on *ptisis*, "sicut gutta aque cavat lapidem racione frequencie, ita stillicidium reumaticum ipsum ulcerat": *Lilium medicine*, 4.4, fol. 51v. This seems taken from Bartholomeus, *De proprietatibus rerum*, 7.31.

[82] "In libro de proprietatibus rerum scribitur quod smaragdus suspensus ad collum curat hemitriteum": *De conservacione*, 2: *De urinis*, 18, fol. 41v. The emerald's properties are given in *De proprietatibus rerum*, 16.88.

[83] See Berthold L. Ullman, "Classical Authors in Certain Medieval Florilegia," *Classical Philology*, 23-27 (1928-1932).

[84] See, for example, the pleas of Isidore of Seville, *Etymologiae*, tr. Sharpe, p. 64.

[85] See Pearl Kibre, "The *Quadrivium* in the Thirteenth-Century Universities (with Special Reference to Paris)," in *Arts libéraux et philosophie au moyen âge*, pp. 175-191.

a 1269 letter of Pope Clement IV to the effect that "it was doubtful whether even arithmetic was taught" there.[86] But Germain may have misread the pope's purely hypothetical assertion that arithmetic "was perhaps (*forsitan*) never lectured on" at Montpellier.[87] In any event, indirect evidence in the writings of physicians who belonged to this university suggests that instruction, especially in arithmetic and astrology, was available, either in the faculty of arts or in that of medicine,[88] but that it may have been taken for granted.

Bernard presumably also took the quadrivial arts for granted, since he did not name them specifically when listing the qualifications which he considered essential for the physician. Among these, he indicated the need for "good visual and sense perception, quick skill, good memory and judgment, and sound intelligence, as Haly Abbas stated."[89] If Bernard was here referring to the Persian Haly Abbas or 'Alī ibn al-'Abbas (d. 994) and his *al-Kitāb al-Malakī* or *Libri X pantechni sive theorice, est liber regalis*, as translated by Constantinus Africanus (ca. 1075),[90] it would be quite remarkable that he omitted the statement that "anyone who wishes to excel in the art of medicine should have a knowledge of dialectic and of the entire quadrivium."[91] However, since several manuscripts read "as Haly stated in the *Tegni*,"[92] it is possible that Bernard was citing the commentary on Galen's *Tegni* or *Microtegni* by the Egyptian Haly or 'Alī ibn Riḍwan (eleventh century).[93] Another requirement, which Bernard

[86] "De quelles sciences pouvait-il s'agir, alors que nous lisons dans une lettre adressée en 1268 [*sic*] par le pape Clément IV au roi d'Aragon Jayme Ier, ... que l'arithmétique elle-même était douteusement enseignée dans nos Écoles?" Germain, *Cartulaire*, introduction, p. 27.

[87] The statement was made in the course of Pope Clement's defense, against James I of Aragon, of the right of the bishop of Maguelone to give the license in civil law, even though the latter had never before laid claim to that prerogative; "sicut hodie servaretur si quis vellet legere arithmeticam, quae nullo forsitan ibi tempore lecta fuit": *Cartulaire*, document 13.

[88] See my "Theory and Practice," pp. 108-109.

[89] Note 94 below.

[90] It was also translated, as *Liber medicine sive theorica, regalis dispositio*, by Stephen of Antioch in 1127: see M. Ullmann, *Die Medizin im Islam* (Leiden, 1970), pp. 140-146. Since Bernard used both titles throughout his writings, he apparently knew both translations.

[91] "Oportet tamen qui in hac arte primatum vult optinere dialectice et tocius quadrivii noticiam habere": *Costantino l'Africano, l'Arte universale della medicina (Pantegni), parte I - libro i*, ed. Marco T. Malato and Umberto de Martini (Rome, 1961), p. 42.

[92] See the variant in n. 94 below.

[93] See *Galieni principis medicorum Microtegni cum commento Hali Rodohan*, in *Articella* (Venice, 1487), fols. 151-210.

emphasized, was that the physician ought to be "able to interpret the sayings of Hippocrates and Galen, and to discern universals and particulars." The physician, Bernard added, "must be knowledgeable (*sciens*)";[94] but how he acquires that knowledge, Bernard unfortunately did not indicate. He further believed that the physician must be diligent and, above all, he must "practise according to the art, that is by the canons of Hippocrates, Galen, and the other sages." These canons, he asserted, "condemn every art of augury, divination and soothsaying, such as the geometric art...."[95] It is obvious that Bernard did not have in mind the traditional mathematical art of geometry. At first glance, one might assume that by the "geometric art" he meant geomancy; but this assumption is questionable unless now, in 1299, he was condemning what he himself had engaged in, a mere four years earlier, when he translated a geomantic treatise.[96] It is more likely that he was referring to an occult science, called "geometrimancy" at a later time by the Aragonese inquisitor Nicholas Eymeric (1320–1399) in his tract *Contra alchymistas*, that is the science described by Professor Thorndike as "the method of divining by chance marking of points or scattering of grains of sand."[97] It is intriguing that Bernard listed the *ars geometrica* among the occult practices and neglected to differentiate it from its quadrivial counterpart. This neglect seems in contrast with his habit of drawing or implying distinctions between rational recipes and occult remedies or between established astrological doctrine and popular superstitions in therapeutics, as we shall see in later chapters, or between the medical and magical applications of "alchemy." As to the latter, for example, he cautioned that "the alchemical method is useful in many ways in

[94] "Preter condiciones, quas ponit Haliabbas (Haly in tegni), scilicet quod sit bonae visionis in quinque sensibus, velocis solertiae, bonae rememorationis, boni iudicii et sani intellectus, oportet quod habeat in se tria, quod sit sapiens (sciens) et diligens, et quod operetur secundum artem. Prima istarum (istarum condicionum) est summe (valde) necessaria, ex hoc quod sit (ex hoc enim erit) potens interpretari dicta (sermones) Hippocratis et Galeni, et universalia a particularibus discernere, universalia (et universalia) particularibus applicare, quare oportet quod sit sciens": Bernard de Gordon, *De decem ingeniis curandorum morborum*, pr. with *Lilium medicine* (Venice, 1498), fol. 1r (at the New York Academy of Medicine). In parentheses are the relevant variants in MSS Cues 308, fol. 106v and Vatican Pal. 1083, fol. 283v.

[95] "Secundo quod sit diligens.... Tertio operetur secundum artem scilicet per canones Galeni et Hippocratis et aliorum sapientium, et per hoc condempnatur omnis ars augurandi, vaticinandi, et divinandi sicut est ars geometrica (geometrialis) et suspendendi herbas ad collum...": ibid. (MS variant in parentheses: see also chapter 5, p. 157, n. 117 below).

[96] See chapter 2, pp. 43-46 below.

[97] Thorndike, *Magic*, 3: 515.

medicine, but in other instances it is so wretched that countless have perished in its practice."[98]

The study of another quadrivial art, arithmetic, was also passed over in silence by Bernard. His proficiency in this art, however, is suggested in his work on medicinal degrees. In his treatise *De gradibus* (1303) he examined the mathematical pharmacy of Galen, Alkindi, and Averroës, and he defended the calculatory techniques of his colleague Arnold of Villanova.[99] In the same essay on the quantification of compound medicines Bernard freely manipulated fractions and proportions which, as he asserted, "we cannot know without numbers."[100] Moreover, Bernard applied arithmetic to another mathematical art, astronomy or astrology. He did this particularly in the calculation of the calendar, the *computus*.[101] At least one fourteenth-century manuscript of a *Ratio computandi per digitos* is attributed to Bernard de Gordon.[102] Even if this should be a spurious work, Bernard did display his familiarity with this art when he discussed the differences between the common, the ecclesiastical, and the medical computations of the lunar month. He further defined in detail the rising and descending quadrants and semiquadrants of the lunar period and asserted that the physician "should have one correct calendar of lunation, for without it he is not a [true] physician."[103] Bernard advised that, not only for prognostication but also for actual treatment, "the physician should know something about the nature of the heavens."[104] As

[98] "Modus alqimicus in multis est utilis in medicina, sed in aliis est ita tristabilis quod in eius via infinitissimi perierunt": *Lilium medicine*, fol. 16r. Bernard's own medical application of alchemy to medicine is evident in his citation of "*alkemiste*" as the source of his information about distilled blood: *De conservacione*, 1: *De flebotomia*, fol. 21v; in the prescription of "thrice distilled human blood" in his own recipe against old age: *De marasmode*, in Vatican MS Pal. 1234, fol. 139v; and in several references to "*aqua ardens*."

[99] See chapter 2, pp. 49-51 below.

[100] "Proportiones autem sine numeris cognoscere non possumus": *De gradibus*, ed. Julius Pagel, *Pharmaceutische Post*, 28 (1895), 182.

[101] Sarton, who preferred the spelling *compotus* (from *componere*) over *computus* (from *computare*), described the *Compotus manualis* of one Magister Anianus, who probably lived near Montpellier in the second half of the thirteenth century: *Introduction*, 2: 992 (with bibliographical references). See also ibid., p. 964 for a *Compotus* ascribed to Roger Bacon.

[102] Cambridge University Library MS Mm.IV.41, fol. 100r, *incipit*, "Hic incipit prima computatio, VI nonas Aprilis signat": see TK 620.

[103] "Sed quia forte medicus ad equacionem omnium planetarum scire non potest, ideo ut possit se excusare tenetur habere tabulam adequacionis lune, quo ad medium motum, qui communiter tenetur habere, et habeatur unum et rectum calendarium lunacionis: sine autem his non est medicus": *Liber pronosticorum*, fol. 101r.

[104] "Medicus debet scire aliqua de naturis superiorum. Ista autem pauca posui non ad

the "something" suggests, however, it seems that Bernard required only a rather basic knowledge of astronomy on the part of the ordinary physician.[105] This impression is borne out by his distinction between the astronomer's and the physician's definitions of the seasons. Spring, for example, "according to the interpretation of the astronomers starts with the ascendancy of Aries and lasts to the setting of Gemini, thus including three signs." The physician's definition, if more poetic, is far less precise or quantitative: "according to the interpretation of the physicians, however, that entire time is vernal in which the trees begin to bud and flower, and the earth begins to germinate."[106]

Bernard's own endeavors, on the other hand, indicate a pervasive interest in the various branches of astronomy. His chief concern was with applied "astronomy" or, more accurately, medical astrology, that is, the influence of the moon and stars on acute diseases, pregnancy, and phlebotomy.[107] He usually did not name specific authorities on this medical astrology, with one single exception where he cited his source for the information that, of those who are born under the sign of Saturn, some are inclined to deviant sexual behavior and "are so evil that the more foul and vile it is, the more they enjoy it, and this is what Haly mentions [in his commentary] on the *Centiloquium*."[108] In all other cases Bernard referred merely to the "wiser astronomers" or to the "Egyptian sages," when he was evidently borrowing from the writings of Albumasar (Abū Ma'shar) or from the above cited pseudo-Ptolemaic *Centiloquium*, especially in his repeated comments on the principle that "the face of the world below depends on the face of the heavenly bodies."[109] It may be mentioned here that Bernard also revealed his

---

alia nisi quod quilibet tenetur scire suo posse naturam superiorum, aliter enim non poterit esse perfectus in sciencia predicendi nec eciam operandi": ibid.

105 See the *aliqua* and *suo posse* in n. 104 and the *ut possit se excusare* in n. 103 above.

106 "Tempus autem vernale secundum intencionem astronomorum incipit a principio arietis et durat usque ad finem geminorum, continetque tria signa.... Secundum tamen intencionem medicorum, [illud] totum tempus est vernale in quo arbores incipiunt pullulare et florere, et terra germinare": *Liber pronosticorum*, 2.1, fol. 80v.

107 For chapters on critical days of acute diseases, see ibid., 5.6: "Docet investigare naturam dierum criticorum per racionem sumptam a quadris lune," and 5.7: "De natura et complexione superiorum quantum pertinet medico"; on pregnancy, see *Lilium medicine*, 7.16: "De difficultate partus"; and on phlebotomy, see *De conservacione*, 1: *De flebotomia*, 14: "Die diebus egyptiacis et kalendis et lunacionis et similibus que considerantur in flebotomia."

108 "Aliqui sunt ita maliciosi quod quanto fetidius est et turpius [sc. coitus in propatulo], tanto magis delectantur, et de hoc facit mencionem Aly supra centiloquium et sunt illi in quorum nativitate dominabitur Saturnus etc.": *Lilium medicine*, fol. 89r.

109 "Vultus inferior est subiectus vultui superiorum..."; of this principle, which is

interest in a more controversial branch of astrology, namely judicial astrology, when he translated a pseudo-Ptolemaic treatise *De reductione geomancie ad orbem*, and possibly also when he wrote a tract on the making of astrological seals and images.[110] Both works, though merely horoscopic, juggle the signs and planets of the zodiac and their properties.

Such discussions of both medical and judicial astrology do indeed suppose a basic working proficiency in physical astronomy or "astrology of motions." This supposition is further borne out in Bernard's case by the instances in which he displayed his familiarity with scientific astronomy. He thus precisely dated his *De conservacione vite humane* as "begun on 22 February 1308 when the moon was in conjunction with the sun towards the seventh degree of Pisces."[111] Moreover, as indicated earlier, he had requested that Profatius Judeus draw up a map of the celestial sphere.[112] Bernard seems even to have tried his hand at astronomical theory when he speculated that the epicycles of planetary motion do not exist in fact but only in a working postulate.[113] Again it may be noted that he did not mention any authority on physical astronomy. In fact, it would appear that his knowledge of astronomy was primarily the result of his own personal interest and investigation, rather than of formal education. It was apparently the achievement of a "hobby" which he cultivated during his career as a physician, rather than the fruit of schooling in the liberal arts.

Also the last of the quadrivial arts, music, appears only as an implicit part of the physician's concern in Bernard's writings. In his above mentioned list of the physician's qualifications. he omitted Haly Abbas's emphasis on the necessity of musical knowledge, which had been cited by such authors as Robert Grosseteste and Roger Bacon.[114] On the other hand, Bernard listed "a perfect knowledge of all musical proportions"

---

found in *Verbum 9* of the *Centiloquium*, Bernard gives some of the applications which the "sapientes egypcii narrant": *Liber pronosticorum*, fols. 100r-101r. For a parallel idea in Aristotle, see n. 122 below. I am indebted to Professor Richard Lemay who has drawn my attention to the pervasive influence of the *Centiloquium* on Bernard's medical astrology.

[110] For a more elaborate discussion of this translation, see chapter 2 below; a treatise on astrological seals and images, attributed to Bernard, is examined in chapter 3 below.

[111] "Inchoatus est enim liber iste in preclaro studio montis pessulani, anno incarnacionis domini M.CCC°vij°xxij° die februarij, luna celebrante synodum cum sole versus vii gradum piscium": Vatican MS Pal. 1174, fol. 10r. This dating is discussed by Ynez O'Neill, "The History of the Publication of Bernard of Gordon's Liber de Conservacione vite humanae," *Sudhoffs Archiv*, 49 (1965), 271.

[112] *De armillis*; see p. 10 above.

[113] "In corporibus supercelestibus, ubi calculans ponit planetas moveri per epicyclos et licet secundum veritatem non sit sic, tamen ad equationem eorum habere non possumus, nisi attribuatur eis motus epicyclorum": *De gradibus*, ed. Pagel, p. 221.

[114] See Kibre, "The *Quadrivium*," p. 186.

among the prerequisites for reading the pulse.[115] He also repeated the traditional belief in the therapeutic value of instrumental music for mental or psychosomatic problems.[116] Furthermore, in the moralizing vein that was one of his salient characteristics, he added a warning against the corrupting influence on children not only of "indecent songs, verses and dances," but also of "evil music."[117] He wondered, too, about the skill with which some musicians pluck an instrument melodiously without thinking, so that their finger moves "by habit, as if it had a mind of its own."[118] Obviously, however, these instances pertain to music as a craft rather than to the study of the mathematical *ars musica*.

The evidence is, indeed, at best inconclusive and at worst negative on the question whether Bernard had received any instruction in the arts of the quadrivium before or during his medical studies. He may possibly have studied part of the arts while already in medical school. At Montpellier in this very period the line between the liberal arts and medicine was not yet clearly drawn. In 1239, for example, Cardinal Guido Soranus, conveying papal confirmation upon the statutes of Montpellier's faculty of medicine, placed "the medical science *among* the liberal arts."[119] It was only in 1340 that the members of the medical faculty expressly enjoined the bachelors from receiving a Master of Arts degree once they had made their public inception in medicine.[120] The

[115] The physician should have "perfectam noticiam omnium proporcionum musicalium," *De conservacione*, 3: *De pulsibus*, fol. 52r.

[116] Thus, for the manic and melancholic patient "debent ibi esse instrumenta musica et breviter omnia que letificant animam": *Lilium medicine*, 2.19, fol. 30v; for lovesickness "dicit Avicenna quod aliqui plus moventur per instrumenta musica": ibid., 2.20, fol. 31r; and for lethargy, ibid., 2.12, fol. 27r.

[117] "[Puer] debet vitare inhonestas cantilenas et versus et ritmos ... dicit Galenus in De regimine sanitatis quod inspectus et auditus et mala musica mores puerorum corrumpunt": *De conservacione*, 4: *Regimen sanitatis*, 5, fol. 57v. Also, "mores puerorum immutantur propter audita et inspecta, et similiter malas musicas": *Liber pronosticorum*, 2.5, fol. 82v.

[118] "Nonne videmus in motibus manuum, quod est mirabile in illis qui tangunt instrumenta, quoniam aliquando ex consuetudine digitus [movet] in melodia et ad cordam et ictum preter hoc quod non intelligat illud quod dirigitur et movetur, quod patet quia aliquando recte viellabit et cum aliquo loquitur. Ecce igitur quomodo digitus ex consuetudine movetur ex se ad locum tamquam si haberet intellectum. Idem dico de pedibus in textoribus": ibid., fol. 83r.

[119] "In firmamento scienciarum inter artes liberales medicinalis sciencia tanquam luminare maius noscitur prenitere": *Cartulaire*, document 4.

[120] "Item, statuimus quod, postquam aliquis Baccalarius promotus suum fecerit principium publice in ecclesia Sancti Firmini in Medicina recipiendo insignia magistratus, non possit in artibus publicam [lauream] nec insignia magistratus recipere": ibid., document 68, art. lxiv.

same statutes forbade the masters "to lecture in the medical schools on any of the textbooks of grammar, logic, or natural science, with the exception of the book *De animalibus*."[121]

Natural philosophy or science, which followed the trivium and quadrivium in the arts curriculum, was inextricably intertwined with the study of medicine. It seems impossible, therefore, to determine how much of Bernard's knowledge of the *philosophia physica* preceded his medical training. While his ideas on natural science will be examined in a later chapter in the context of medical theory, we may briefly indicate here some possible evidence of his schooling in that branch of philosophy. Such evidence could perhaps be found in his frequent citations not only of Aristotle's *De animalibus*, but also of the *De anima*, *Historia animalium*, *De motibus animalium*, *De coloribus*, and the *Metaphysica*. In numerous cases Bernard cited the specific chapter in these Aristotelian works.[122] Furthermore, he appeared quite familiar with such elements of Aristotelian ontology as *potentia* and *actus*, and *materia* and *forma*, although he may have acquired this familiarity either through purely medical channels, such as the writings of Galen and Avicenna, or through scholastic commentaries. It seems surprising, however, that he named only two modern authorities on natural science, namely Alveredus or Alfred of Sareshel[123] and Albertus Magnus. Of the latter author, the only one whom he called *dominus*, Bernard cited the treatise *De nutribili et nutrito*[124] and the *Phisica* or *Philosophia*,[125] which was Albert's introduction to Aristotle's *Physics*.

The last of the three philosophies, which made up the arts curriculum,

[121] "Item, statuimus quod nullus magister legat vel permictat legere in scolis medicinalibus aliquem librorum in cathedra grammaticalis vel logicalis vel naturalis sciencie, ullo modo, excepto libro 'De animalibus'": ibid., art. xxxi. By *De animalibus* was meant Avicenna's version of Aristotle's *De generatione animalium*.

[122] In addition, Bernard appears to quote Aristotle's *De caelo et mundo* in his statement that the "radix superior adheret vie uni et ordini uni, sicut sunt motus orbium": *De conservacione*, 4: *Regimen sanitatis*, prol. fol. 53v (but Bernard here also refers to Galen's *De criticis diebus*).

[123] "Alvaredus scripsit quod cor est vite domicilium et vite consulens": *De conservacione*, 3: *De pulsibus*, fol. 51v. The reference is probably to Alfred's *De motu cordis* (ca. 1210), for which see TK 96 and 1020; Thorndike, *Magic*, 2: 195; and Sarton, *Introduction*, 2: 561.

[124] "Omnem potenciam precedat actus secundum intencionem domini Alberti in De nutribili et nutrito": *Lilium medicine*, 7.14, fol. 95r. For Albert's *De nutrimento et nutribili* in MSS, see TK 98 and 365.

[125] The *vis imaginativa* never rests "secundum intencionem domini Alberti in phisica sua": *Lilium medicine*, 2.15, fol. 28v. In the incunabulum *Lilium* (Venice, 1498), fol. 29r, we read "in philosophia sua."

was moral philosophy. This branch of learning ranked especially high in some writings of Bernard. He utilized the nonscientific authorities almost exclusively for moralizing purposes. Thus, more than half of his quotations from Aristotle relate to ethical subjects. Aristotle's *Ethics*, which was an important textbook in the arts curriculum of some universities,[126] is cited at least fifteen times. Furthermore, the Latin classical authors were summoned primarily to buttress Bernard's pious admonitions. This should not really surprise us, because since the twelfth century the pagan Latin authors were called "*ethici*" or moralists,[127] and in some cases ethics was part of the grammar curriculum.[128] It was especially to Seneca, who "spoke with an angelic tongue,"[129] that Bernard turned for ascetic advice against self-indulgence and, interestingly enough, against the pursuit of "curiosity" in studying.[130] In fact, it is worth noting that the guidelines, which he drew up for the student, the master, and instruction, heavily emphasized ascetic prescriptions and contained only few intellectual criteria. Bernard also called upon the Fathers and more recent spiritual authors to underscore his pleas for a virtuous life of moderation, chastity, and humility.

The character of these guidelines, when combined with the authors chosen to underscore them, invites one to conjecture that Bernard may have received his earliest education in monastic circles. To be sure, any liberal arts background might account for his knowledge of Saint Jerome and Saint Augustine, as well as of Hugh of Saint Victor. However, his references to Prosper of Aquitaine and to Saint Bernard of Clairvaux seem to point towards the literature of the cloister. It is especially remarkable that he cites Prosper's *De vita contemplativa* which, unlike his *Epigrams*, seemed not to belong to the traditional curriculum,[131] and also that

[126] Thus, at Toulouse in 1309: Paetow, *The Arts Course*, pp. 96-97.

[127] For example, Gilbert de la Porrée (in Migne PL 164: 1316); Hugh of St. Victor (PL 175: 10). Peter Abailard called Seneca "ille maximus morum aedificator": *Dialogus*, ed. Victor Cousin *Petri Abaelardi Opera*, 2 (Paris, 1849), 675.

[128] J. Reginald O'Donnell, "The Liberal Arts in the 12th Century with Special Reference to Alexander Nequam (1157-1217)," in *Arts libéraux et philosophie au moyen âge*, p. 130.

[129] "De qualitate cibi et potus ... loquitur Seneca lingua angelica": *De conservacione*, 4: *Regimen sanitatis*, fol. 59v.

[130] "[Discipulus] non studeat in curiositatibus ubi modica aut nulla est utilitas et labor magnus, et huius curiosi sunt qui student in profetiis inveniendis et similibus, et de hoc Seneca, Quare te torques in illa questione quam subtilius est contendisse quam solvere": ibid., fol. 59r.

[131] See Curtius, *European Literature*, passim, and R. R. Bolgar, *The Classical Heritage* (New York, 1964), p. 257.

Bernard invokes that work on contemplation to support not one of his moralistic but one of his few "academic" guidelines, namely that the master's teaching should be comprehensible to even the least bright students.[132] Master Bernard had perhaps begun his schooling under the Cistercians, who founded an abbey at Gourdon in 1261.[133] In any event, he was a deeply pious man, one who, at least in the closing years of his life, believed with Saint Augustine that "doubts are solved better by prayer than by other inquiry."[134] Prayer, Bernard held, was also an effective soporific: he advised the insomniac patient, for whom all other remedies for sleeplessness failed, to "start saying the Divine Office, and he will sleep immediately."[135] In addition to referring to these monastic prayers, he also echoed the asceticism of monastic practices when he prescribed fast and scourging against concupiscence.[136]

The possible influence of monastic asceticism upon Bernard is further suggested by some aspects of his expressed attitude towards woman, love, and marriage. He held that "looking at a woman causes the worst corruption of morals,"[137] and he called love a "disease (*insania*) of the mind, that mixes a few pleasures with its sorrows."[138] While views like these pervaded the biblical, classical, and patristic literature and therefore were hardly original with Bernard, their application in his writings points to a personal attitude rather than simply to a passive adherence to cultural-social prejudices. One striking illustration is found in his peculiar application of Saint Paul's well-known concession that "it is better to marry than to burn" (1 Cor. 7.9). In his chapter on "*gomorrea*" (*sic*, thus etymologically relating gonorrhea and Gomorra), which he defined as "involuntary emission of sperm," he applied Saint Paul's dictum to the

[132] With Aristotle's *Metaphysics* "concordat Prosper Aquitaneus in libro de vita contemplativa cum apponit, Debet esse sermo docentis ut ab intelligencia sua nullos excludat quamvis imperitos": *De conservacione*, 4: *Regimen sanitatis*, fol. 59r.

[133] Chevalier, *Répertoire*, Vol. 1: *Topo-Bibliographie*, col. 1326. The date of the Cistercian foundation is given as 1241 by Pierre Deschamps, *Dictionnaire*, p. 574.

[134] "[Discipulus] libenter et frequenter oret et de hoc Augustinus, Oracione melius solvuntur dubia quam inquisicione alia": *De conservacione*, 4: *Regimen sanitatis*, fol. 59r.

[135] "Incipiat dicere horas dominicas, et statim dormet": *Lilium medicine*, 2.18, fol. 29v.

[136] "[Contra concupiscenciam] primo inducantur ieiunia, accidencia terribilia et flagella": ibid., 7.3, fol. 90v.

[137] "Aspectus autem mulieris inter omnia corrumpit summe mores": *De conservacione*, 4: *Regimen sanitatis*, fol. 53v. Bernard's attitude towards woman is discussed briefly in my article, "The Idea of Childhood and Child Care in Medical Writings," *Journal of Psychohistory*, 4 (1977), 481-483.

[138] "Amor est mentis insania, qua animus vagatur per inaniam cerebri, doloribus permiscens dulcia (Venice 1498: pauca) gaudia": *Lilium medicine*, 2.20, fol. 31v.

advice that marriage may be a last resort when *gomorrea* is caused, not by concupiscence, but by an overabundance of sperm.[139] At the same time, however, Bernard believed that the presence of women might be bad for a man's health. He blamed the death of a colleague, apparently of a stroke, on the fact that he had been stuffed too much with ground chicken by women, "since one can hardly follow a good regimen where women wield power."[140] It also seems strange, and it is perhaps indicative of his attitude towards marriage and women, that Bernard referred to the newborn infant as "*baccalaureus*":[141] even though it could be translated as "a young lad," this designation is a subtle deviation from the usage of the time, in which the term bachelor commonly had the same social and academic meanings as it has in English today. A clue to Bernard's own bachelorhood may possibly be found in his recommendation of an aphrodisiac whose effectiveness he presumably knew only from hearsay, when against impotence he prescribed a "concoction of which good colleagues say that it makes the woman faint under the man."[142]

Although he seems to have been celibate, it is not likely that Bernard was a professed monk. He did esteem the religious life, but sometimes for rather unusual reasons. For example, he appreciated "the peace of mind and the withdrawal from worldly affairs," because it gave the religious a better memory.[143] In his explanation of a related aspect, monastic silence, he was equally mundane in conjecturing that "one reason why the religious do not talk at table" may be so that they would neither choke on their food nor swallow air.[144] He also devoted special consideration to monks in his remedies, such as one for improved sight that "enables lectors and monks to read small letters."[145] However, when he prescribed a special ointment against lice that was "suitable for the religious," we detect a less admiring note as he elaborates that, like the poor and the "wild men who care little for the hygiene of their body," the religious also

[139] *Lilium medicine*, 7.3, fol. 90v. Bernard's remedy for concupiscence is in n. 136 above.

[140] "Gallinis tritis multum replebatur a mulieribus, quia vix potest aliquis bene dietari ubi mulieres habent potestatem": *Lilium medicine*, 2.12, fol. 27v.

[141] *De conservacione*, 4: *Regimen sanitatis*, passim.

[142] "Deinde utatur ista decoctione de qua dicunt boni socii quod facit mulierem deficere sub viro": *Lilium medicine*, 7.1, fol. 89r.

[143] "Religiosi bene memorantur propter quietem anime et dimissionem negociorum secularium": ibid., 2.12, fol. 27v.

[144] "Non est bonum cum aliquis loquitur quod comedat quia cibus transiret ad pectus nec e converso quia aer transiret ad stomachum, et hoc potest esse una racio quare religiosi non loquuntur in mensa, ne una operacio impediat aliam": ibid., 3.23, fol. 46r.

[145] "Valet lectoribus et monachis ut sciant legere literas minutas": ibid., 7.21, fol. 98v.

"teem with lice."[146] In fact, he appeared to share some of the laity's less charitable opinions of the religious. Thus, in an interesting description of depressive mania, he listed among the symptoms the nightmares in which the patient saw such "frightful things" as devils, hanged dead, and black monks.[147] He may also have included, as possibly dispensable, the religious in his medical experiments. He advised, namely, that the strength of a medicine should be tested "first on birds, next on mammals, then [on patients] in hospitals, then on lesser brethren, and then on others in [ascending] order, because if it should be poisonous it would kill."[148] If the phrase "*in fratribus minoribus*" referred to the lower members of society in general, it strangely blurred the contemporary distinction between the class of *homines minores* and the religious order of mendicants. This oversight or, a fortiori, the intended inclusion of the Friars Minor among the human guinea pigs would suggest that Bernard himself was not a Franciscan and that, unless he was a forgetful or vengeful alumnus, he had not received his education in that order.

The process of elimination and conjecture, that has characterized the above attempts to trace Bernard's intellectual background, still prevails when we explore the setting of his medical education. Here, however, we have more direct help in his own statements and in the colophons of his writings. The content and nature of his works and thought will be examined in detail in the following chapters with attention to the context of medical teaching at Montpellier. With regard to his training in medicine, as on the subject of his national origin, assumptions have been made that have no basis in known facts. Thus, George Sarton guessed that Bernard "was educated at Salerno,"[149] probably because he displayed an extensive knowledge of the Salernitan works and doctrines. However, the Montpellier faculty of medicine itself in the thirteenth century generally cultivated the Salernitan tradition, as may be seen in the university statutes and in the writings of its earlier members. In addition, the absence of

[146] "Rx ... et est congruum religiosis.... Pauperes enim et religiosi, qui multiplicant cibaria flegmatica et melancolica, habundant in pediculis, similiter homines ferales qui de cultu sui corporis modicum curant": ibid., 2.9, fol. 23v.

[147] "[Paciens] habet fantasmata diversa et terribilia quia videtur sibi in sompno quod videat demones aut monachos nigros aut suspensos et mortuos": ibid., 2.18, fol. 30r; also *Liber pronosticorum*, 1.3, fol. 75v.

[148] "Si aliquam medicinam volumus experiri in corpore humano, quod prius experiamur in avibus et postea in brutis animalibus et postea in hospitalibus et postea in fratribus minoribus et postea in aliis per ordinem, quia si forte esset de genere venenorum interficeret": *De gradibus*, ed. Pagel, p. 143.

[149] *Introduction*, 3: 873.

personal reminiscences about southern Italy, which we have emphasized earlier, suggests that it is unlikely that he ever visited Salerno. There is, on the other hand, more substantial evidence in one of his own treatises, the *Compendium regiminis acutorum*. He had, as he asserts in the preceding treatise *De regimine acutorum*, originally composed the synopsis "while still a bachelor, at the time of the first lecture (*cursus primus*)."[150] If this baccalaureate lecture was delivered at Montpellier, as we may presume from the absence of any indications to the contrary, it provides a clinching argument that Bernard had received his medical education there.

A dual frustration faces the student of intellectual history who wishes to situate Bernard de Gordon in a living and personalized sequence of medical teaching at Montpellier. In the first place, Bernard occasionally referred to "our masters" but failed to name or otherwise identify a single one of them, even though every student had to be "assigned to the direction of one certain master" (*certi magistri addictus regimini*) according to the original university statutes.[151] In addition, the period between 1260 and 1285, which must have embraced the years of Bernard's progress to master in Montpellier's faculty of medicine, has not only been studied less thoroughly by historians but is also less richly documented in the official records and scholarly writings than the preceding era. Especially the first decades of the thirteenth century had been full of activity for the budding medical faculty, which was granted its first statutes in 1220.[152] In this period the champion of Salernitan learning, Aegidius Corboliensis or Gilles de Corbeil (ca. 1140–1224) had come to the town but, perhaps not finding sufficient adulation, had turned his back on the "*Mons Pessulanicus errans*."[153] A few years later the surgeon Guillelmus de Congenis (ca. 1215) had operated in the Hospital of the Holy Spirit in front of his students with the express intention that "they

[150] The *Regimen acutorum* ends, "hec autem omnia scripta sunt post lecturam continuam xi annorum. Qui autem ista brevius habere voluerit accipiat subsequencia que tamen fuerunt facta in statu baccalarii tempore cursus primi etc. Explicit tractatus unus de regimine acutorum bernardi de gordonio quem composuit post lecturam xi annorum. Sequitur alius de eodem eiusdem adhuc in baccalariatu existentis," and here follows the *Compendium regiminis acutorum* in Vatican MS Pal. 1083, fol. 281v.

[151] *Cartulaire*, document 2.

[152] Ibid.

[153] Gilles de Corbeil, *De urinis*, v. 348, cited by Ernest Wickersheimer, "La question du Judéo-Arabisme à Montpellier," *Monspeliensis Hippocrates*, 6 (1959), 6. The medical poems of Gilles were ed. by Louis Choulant, *Carmina medica* (Leipzig, 1826). See also Gernot Rath, "Gilles de Corbeil as Critic of his Age," *Bulletin of the History of Medicine*, 38 (1964), 133-138, and further references there.

might learn to operate not only by hearing but also by seeing."[154] By the middle of the century the flourishing university boasted a diversity of masters, ranging from the somewhat pedantic Gualterus Agulinus (Walter Agilon), who built most of his medical theory and practice around uroscopy,[155] to Chancellor Henricus de Guintonia (Henry of Winchester) and his mysteriously named colleague Magister Cardinalis, a scholarly pair who are becoming better known thanks to the careful study of their writings by Professor McVaugh.[156]

As may be noted, the above named masters were still almost a half century removed from Bernard's formative period in the medical faculty. He probably also came too late for the next generation that included Rogerius de Barone (fl. 1260), who composed a four-part *Practica* known as *Rogerina*,[157] and the empiric Nicholas de Polonia (fl. ca. 1275), who wrote an iconoclastic tract titled *Antipocras*.[158] Around that time the medical university had begun to act more vigorously against practice by unlicensed persons "of both sexes, Christians and Jews," who threatened not only the common welfare but also the "name and fame of the *Studium*." These fears were expressed, and the term *licentia* was apparently used for the first time, in an edict that James I of Aragon issued in 1272 at the request of the members of the faculty of medicine.[159] The year 1281, in which Bernard was presumably preparing for his baccalaureate, was marked by further diplomatic efforts of the masters to protect their guild interests. The *Cartulaire* contains several notarized letters which they addressed to the official of the bishop of Maguelone. Thus, on 20 March the masters complained that one Johannes de Chipro had practised medicine in violation of "the tenor and form" of their privilege, and they claimed restitution for the "expenses and damages" which they had suffered as a result of that practice.[160] The financial aspect of their complaint underscores the concrete character of the masters' guild

[154] Cited by Karl Sudhoff, "Gedanken über die Ausbildung chirurgischer Operateure an den Hochschulen im 13. Jahrhundert," *Mitteilungen zur Geschichte der Medizin*, 17 (1918), 294.

[155] Gualterius Agulinus, *Summa medicinalis*, ed. with notes by Paul Diepgen (Leipzig, 1911).

[156] Michael McVaugh, "An Early Discussion of Medicinal Degrees at Montpellier by Henry of Winchester," *Bulletin of the History of Medicine*, 49 (1975), 57-71; see also the other studies by Professor McVaugh in the bibliography below.

[157] I have consulted the early Bruges MS 470, fols. 41r-94r.

[158] Ed. Karl Sudhoff, *Archiv für Geschichte der Medizin*, 9 (1916), 31-52.

[159] *Cartulaire*, document 14.

[160] Ibid., document 15.

solidarity.[161] Four days later they protested further that they did not intend to renounce their privileges, "either tacitly or expressly, in word or deed"; nor would they give up any of the stipulations which regarded their "interest and welfare" (*utilitatem et commodum*).[162] On 31 March, moreover, the masters objected to the fact that the official had intervened in examinations without duly showing his authorization; and they demanded that he should "observe, without breach, towards us and towards the said university, our privileges and statutes."[163] It is worth pointing out that, of the masters pursuing the university's and their own interests in these letters of 1281, no scholarly writings have come to us. The letters have been examined briefly here because they vividly reflect some human facets of that academe where Bernard de Gordon was then receiving his training and in which he would spend his entire career as practitioner and professor of medicine.

For information about Bernard's career and biographical dates we turn to his own writings. From the prologue to his *Lilium medicine*, which he wrote in 1303 "after our twentieth year of lecturing,"[164] we learn that he began to teach in the Montpellier faculty of medicine in 1283. If we estimate that he was twenty-five years old at the time of his inception, the date of his birth could be put around 1258. By the same reckoning, he would have been thirty-six when he composed his presumably first work, the treatise *De regimine acutorum* [*morborum*] "after eleven years of continuous lecturing."[165] Most of his other writings are also dated, either relatively as the one just mentioned, or absolutely as the last dated work, the *Liber de conservacione vite humane*, which was composed in Montpellier in 1308. Even though he may have written some other treatises after that year, as will be suggested in a later chapter, there is no reason to believe that he died outside Montpellier. The date of his death has traditionally been set around 1318 or 1320, unfortunately without documentary justification.[166] The earliest confirmation of his death may

[161] The same financial aspect was again underscored in 1319, when Johannes de Lodova, bachelor of medicine, was sued for practising contrary to the statutes, thereby "detrahendo magistris lucra": ibid., document 35.

[162] Ibid., document 15.

[163] Ibid.

[164] "Inchoatus autem est liber iste ... in preclaro studio montis pessulani post annum xx. lecture nostre, anno domini m.ccc. tercio mense julii": *Lilium medicine*, fol. 1r.

[165] "Hec autem omnia scripta sunt post lecturam continuam xi annorum": *De regimine acutorum morborum*, in Vatican MS Pal. 1083, fol. 281v.

[166] "Ranchin [?] le place à l'an 1318; mais, comme il ne cite aucun document à l'appui de son dire, on peut croire qu'il n'y a là qu'une date approximative," Littré, *HL*, 25:

be a notarial document dated 1330. This document concerned Gulielmus Gordonius, a physician and canon of Asti (Monferrato, Italy), and made mention of "the other Bernard Gordonius, *formerly* outstanding teacher of medicine in Montpellier," who lived "in the area of Provence."[167]

Since most of Bernard's works are identified as having been completed "*in preclaro studio Montis Pessulani*," it is safe to assume that he spent his entire teaching career, which spanned at least a quarter of a century, at the same university. This steadfast commitment distinguishes him from several of his more famous contemporaries, who left the faculty after a few years. Pierre Pansier, an historian who was especially familiar with Montpellier around 1300, observed that the service of professor at the school was "not very lucrative" and that the successful doctors had only one goal, "to take advantage of their assignment to gain renown and access to the courts of the great."[168] Pansier's observation is borne out by the careers of two of the luminaries usually considered the leading representatives of Montpellier. Arnold of Villanova (ca. 1240–1311) probably spent as much time at the courts of Aragon and Avignon as at the university; and Henri de Mondeville (d. ca. 1325) became the personal surgeon of Philip the Fair in 1302. Bernard, on the contrary, seems not to have served as physician to potentates, although he may have known someone at the court of King Jayme of Mallorca who brought him knotgrass from the king's herb garden.[169] He in fact recounted only one case in which he treated a noble patient, a certain leprous countess,[170] and only occasionally gave special attention to the aristocracy, as, for example, when he remarked that if a physician should have bad breath, "it would be embarrassing, especially when he converses with nobles."[171]

---

322. Bernard died "vers 1320" according to Cyr U. Chevalier, *Répertoire des sources historiques du moyen âge; Bio-Bibliographie*, 1 (Paris, 1905), 556.

[167] One of the documents for 1330 is "una carta di procura v'è accennato d'un Bernardo del fu *alterius Bernardi Gordonii olim in monte persulato* [sic] *medicine lectoris preclarissimi* cugino di Guglielmo, abitante *in partibus provincie*, il quel Bernardo più giovane protesta di volere ed intendere, che quanto verrà fatto in questa occasione dal suo cugino sia come se fosse fatto da lui medesimo": Vincenzo Malacarne, *Monumenti delle opere de'medici e de cerusici che naquero o fiorirono prima del secolo XVI negli Stati della Real Casa di Savoia*, 2 (Turin, 1789), 92-93.

[168] Pierre Pansier, "Arnaldi de Villanova *Libellus regiminis de confortatione visus*," *Mémoires de l'Académie de Vaucluse* (Série 2), 3 (1903), 2-3.

[169] "Ego feci apportari [poligonia] de viridario serenissimi regis maioricarum": *Lilium medicine*, 3.3, fol. 38v.

[170] "Quedam comitissa venit [leprosa] ad montem pessulanum, et fuit in fine in cura mea": ibid., 1.22, fol. 13r.

[171] "Quoniam si medicus talia pateretur, valde esset ignominiosum, et potissime quando cum nobilibus conversatur": ibid., 3.15, fol. 43v. Compare pp. 156-157 below.

If there is little indication that Bernard consorted with the aristocracy, there is scarcely more evidence that he played an active role in academic politics. In some manuscripts he is referred to as "dean of the physicians" or as "dean of the distinguished *Studium* of Montpellier."[172] The dean's prerogative, as formulated in the 1242 statutes, was to determine the time and duration of recess from lectures and disputations, and to initiate that recess simply by his own adjournment.[173] With regard to the term *Studium*, it is of interest to note that it was during the early years of Bernard's career, in 1289, that the Montpellier schools of arts, medicine and law received the status of *Studium generale*.[174] The manuscript references to Bernard as dean, while possibly accurate, may account for two apparently incorrect characterizations of his university office. Jean Astruc designated him "rector of the college,"[175] although the title of rector was replaced after 1242 by that of chancellor and the term "college" does not apply to the faculty of medicine in the decades around 1300. A more recent historian gave Bernard, as dean, a crucial role in an academic dispute in 1313,[176] evidently confusing him with Bernardus de Bonahora. This Bernardus the records mention as the master (and not the dean) involved in a 1313 provision which required legitimate birth for promotion to master and which led to a dispute in 1315.[177] It is, in fact, striking that Bernard de Gordon's name is totally absent from the group of some twenty-two masters who are mentioned in the documents, preserved in the *Cartulaire*, for the period between 1283 and 1313. The absence of his name seems especially surprising in the preamble to the book list and curriculum requirements drawn up by Pope Clement v at the request of the members of the university and with the advice of two physicians, Arnold of Villanova and Jean d'Alais, "who formerly and for

[172] "Incipit tractatus de gradibus ordinatus a magistro bernardo de gordonio, decano medicorum in almo studio montis pessulani anno domini m.ccc.iij," in MSS Erfurt Amplonianum F. 236, fol. 50v, and Q. 227, fol. 304, both 14th-century, cited by Karl Sudhoff, "Zum Schriftstellerei des Bernhards von Gordon und deren zeitlicher Folge," *Archiv für Geschichte der Medizin*, 10 (1917), 179-180. Also in the possibly earlier MSS of the same treatise, Cues 308, fol. 108v and Paris BN 16189, fol. 195v; and the later Vatican MS Pal. 1083, AD 1458, fol. 321v: "est finis tractatus de sterilitate editus a bernardo de gordonio decano preclari studii montispessulani."

[173] *Cartulaire*, document 6.

[174] Ibid., document 20.

[175] Astruc, "Bernardus Gordonius," p. 181.

[176] Sonoma Cooper, "The Medical School of Montpellier in the Fourteenth Century," *Annals of Medical History* (New Series), 2 (1930), 173. Yet the same author also states that Bernard "was not reading in the school at the time."

[177] *Cartulaire*, documents 32 and 51; see also document 31.

a long time conducted classes there," as well as of "some other masters who presently hold a chair in the aforesaid *Studium*, and of numerous others who, promoted to masters in the same place, conducted classes there at diverse times."[178] Bernard remained nameless among those masters when the pope issued his academic directives in 1309, even though by that time he had produced an impressive *corpus* of writings that were closely related to the Montpellier curriculum.

It is conceivable that Bernard aroused personal resentment in academic circles because of his criticism of colleagues and of contemporary practices.[179] For example, as Professor McVaugh has suggested, Bernard may have criticized the visionary excesses of his mercurial colleague, Arnold of Villanova,[180] even though he deferred to Arnold as one of the founders of the difficult science of quantifying medicinal degrees.[181] Arnold in return, though not naming Bernard in his own works, may have had a running disagreement with him on the subject of practical compendia. The prolific Arnold presumably never wrote a compendium of that type but condemned those who preferred to study in "booklets and voluminous *summas*."[182] Bernard, on the other hand, in the prologue to his *summa* on practice or *Lilium medicine*, banned from his "common table" those snobs (*superbi*) who disdained such elementary writings.[183] If these reciprocal allusions implied a certain dissonance, a note of irony and paradox may be found in the fact that Arnold himself apparentely owned a copy of Bernard's *Lilium medicine*.[184] Bernard was also frequently critical of his fellow physicians in general. Thus, he berated them for their neglect of diagnosis by phlebotomy,[185] as well as for their "arrogance and

[178] Ibid., documents 25-27.

[179] For more on Bernard's critical attitudes see chapter 4 below.

[180] Michael McVaugh, "Nota sobre las relaciones entre dos maestros de Montpellier: Arnau de Vilanova y Bernardo Gordon," *Asclepio*, 25 (1973), 333-335. Also Bernard's criticism of those "qui student in profetiis inveniendis," cited in n. 130 above, may be an allusion to Arnold.

[181] See the discussion of Bernard's *Tractatus de gradibus* in chapter 2 below.

[182] "Potius student in cartapellis et summis que potissime magni voluminis sunt...": Arnoldus de Villanova, *De considerationibus operis medicine*, in his *Opera* (Lyon, 1509), fol. 90v.

[183] "Et quia humilibus (de humilibus) scribo, ex hoc superbi abiecti sunt, separatum enim est convivium eorum; superbi enim non sedent in mensa communi, cum aliis (cum a nobis) scripta communia contempnunt...": *Lilium medicine*, in Paris MS BN 6964, fol. 1r (variants in London MS BL Harl. 3698, fol. 1r).

[184] McVaugh, "Nota," p. 332.

[185] "Medici ut plurimum sunt negligentes in aspiciendo sanguinem extractum per flebotomiam": *De conservacione*, 1: *De flebotomia*, 8, fol. 20r.

ignorance" in spurning uroscopy.[186] Furthermore, he accused them of venality when he charged that they were more interested in the lucrative treatment of diseases than in the preventive application of a good regimen.[187] On one occasion, he even seemed to insinuate that the masters of Montpellier were partly responsible for the death of a colleague. He related the case of a forty-year-old master who had lapsed into a coma: a group of masters had agreed that the patient should be phlebotomized, but not one of them supervised the actual venesection by a surgeon, and the master bled to death.[188]

Insinuations and criticisms like these probably did not endear Bernard as a colleague to his contemporaries at the university. That may in part explain why so little external evidence is available on his character and personal life, even though his teaching was remembered at Montpellier throughout the fourteenth century and his works were soon cited and copied all over Europe. Henri de Mondeville compiled his *Cyrurgia* in 1306 presumably "at the request and command of Bernard de Gordon, man of science and eminent professor of medicine."[189] In 1325 Master Jordanus de Turre identified Bernard as one of "our predecessors" at Montpellier,[190] and in 1365 Johannes de Tornamira still included him among "our modern Montpellierains."[191] At the end of the century Bernard's writings were listed by the German Heinrich Lamme as

[186] "Patet arrogancia et ignorancia quorumdam qui dedignantur et nauseant iudicia urinarum": *De conservacione*, 2: *De urinis*, 1, fol. 28r.

[187] "Ymmo quod est magis abhominabile, medici nostri temporis non curant scire regimen sanitatis quia videtur eis quod non sit ibi lucrum sed in curacione morborum et potissime febrium sunt bene solliciti": *De conservacione*, 4: *Regimen sanitatis*, 17, fol. 72v.

[188] "Fuit mortuus magister noster ex litargia et erat xl (lx Paris BN 6964) annorum, colericus adustus, et erat ibi turba magistrorum, et omnes convenerunt quod aperirentur emorroydes; et cyrurgicus nullo magistro existente aperuit et extraxit ut dicitur sanguinem in nimia quantitate, intantum quod semper processit debilitando donec mortuus est": *Lilium medicine*, 2.12, fol. 27v.

[189] "ad peticionem et preceptum scientifici viri Magistri Bernardi de Gordonio, in preclarissimo studio Montispessulani summi professoris in sciencia medicine": quoted from Paris MS BN 7131 by E. Nicaise, tr. and ed., *La Chirurgie de maître Henri de Mondeville* (Paris, 1893), p. 1. However, it was Guillelmus de Brescia rather than Bernard who issued this request, according to Wickersheimer, *Dictionnaire*, 1: 230.

[190] In the prologue to his *De adinventione graduum in medicinis simplicibus et compositis*, in Vatican MS Vat. lat. 2225, fol. 53ra, brought to my attention by Professor Michael McVaugh. The date of composition is given as 1325 by Pierre Pansier, "Les maîtres de la Faculté de Médecine de Montpellier au Moyen Âge," *Janus*, 10 (1905), 3.

[191] In his *Clarificatorium*, a commentary on the ninth book of Rasis' *Ad Almansorem*, according to Pierre Pansier, "Jean de Tournemire; Étude bio-bibliographique," *Monspeliensis Hippocrates*, 42 (1968), 9.

indispensable in the physician's education,[192] and his views were adopted by the Portuguese Valescus de Tharanta as eminently authoritative.[193] The earliest as well as the later references indicate that Bernard's reputation, whether celebrated at Montpellier or broadcast across Europe,[194] was based on his scholarship rather than on personal panache, academic activism, or association with magnates. For Bernard de Gordon himself, it would appear, his career as author, teacher, and practitioner of medicine was sufficiently engrossing to encompass the whole of his attention.

[192] Ernest Wickersheimer, "Le discours de réception d'un bachelier en médecine montpelliérain au début du xve siècle," *Bulletin de la Société française d'histoire de la médecine*, 13 (1910), 247.

[193] For example, "ego teneo cum Gordonio," *Practica que Philonium dicitur* (no place given, 1500), p. 15v, and similar expressions, passim.

[194] Not only to England, for which see p. 1 above, but also to Germany and Italy, as attested by the provenance of several 14th-century manuscripts of Bernard's works.

2

Dated Works

The most complete bibliographical study of Bernard de Gordon's writings was published in 1917 by the noted historian of medicine, Dr. Karl Sudhoff.[1] On the basis of manuscripts Dr. Sudhoff established the chronology of Bernard's works but, because of World War I, his research was restricted to German libraries. His article, moreover, which brought up to date the pioneer study by Baron von Töply,[2] has itself become out of date in view of more recent manuscript studies and the publication of such valuable guides as Dr. Ernest Wickersheimer's *Dictionnaire biographique*, Thorndike and Kibre's *Catalogue of Incipits*, and so on.[3] These have supplemented Sudhoff's earlier work and have pointed up the need for a bibliographical study that will broaden and update his findings. Such a study should serve also as the foundation of a more comprehensive and more accurate evaluation of Bernard's teaching and personality.

The Gordonian writings fall into two groups, one comprising works about whose authorship there is no reasonable doubt, the other comprising treatises that have been attributed to Bernard.[4] The first group can be treated in chronological order since, with only two exceptions which will be examined in the following chapter, these works are dated, either relatively or absolutely, in the manuscripts. A relative date is given for Bernard's first published tract,[5] *De regimine acutorum morborum*, which

[1] Karl Sudhoff, "Zur Schriftstellerei Bernhards von Gordon und deren zeitlicher Folge; eine Handschriftenstudie," *Archiv für Geschichte der Medizin*, 10 (1917), 162-188.

[2] Robert von Töply, "Die chronologische Reihenfolge der Schriften des Bernhard Gordon," *Mitteilungen zur Geschichte der Medizin und der Naturwissenschaften*, 6 (1907), 94-95.

[3] In Lynn Thorndike, *History of Magic and Experimental Science*, 2 (New York, 1923), passim, the passing references to Bernard are based primarily on the article by Littré in the *HL*, 25: 321-337, and on printed editions of Bernard's works.

[4] The manuscripts of the works composed by or attributed to Bernard are listed in the appendix to this study.

[5] By "published" (*editus*) is meant the public delivery of the treatise, either in a lecture by its author or in a display by the *scriptorium*.

he composed "after eleven years of continuous lecturing," that is in 1294.[6] This brief treatise appears based on the Hippocratic *Regimen of Acute Diseases*, which was part of the curriculum at Montpellier.[7] Bernard's avowed purpose with this work was to set forth the general principles of the treatment of critical or acute illnesses, while leaving the "wise" and "inventive" physician to construct on this framework his own regimen for each individual case.[8] The treatise comprises three parts, of which the first contains a description of the "correct administration of food and drink, with regard to quantity, quality, and sequence."[9] The second part relates to the evacuation of the fever-causing agents by phlebotomy and by such cathartics as rhubarb and black hellebore. The third and last part contains the "digestives which Hippocrates lists, namely wine, hydromel, oxymel, water, and baths."[10] While the dependence of this regimen on its Hippocratic model is evident, Bernard seems to have made a special effort not only to organize the subject matter more logically than it was in the Hippocratic regimen, but also to document it with references to the later medical tradition. In addition to Hippocrates, he also drew on other authorities, such as Galen, Avicenna, and Averroës, although he did not always name his sources. Thus, Hippocrates is referred to only three times, Galen twice, and Avicenna and Averroës once each. On the other hand, Bernard also gave credit to "old wives," from whom he had learned a particular remedy; however, they had taught him "more by way of empirical knowledge (*per viam experimenti*) than by reason."[11] He drew attention, moreover, to instances or cases in which the physician might err either because of insufficient insight or because of lack of anatomical knowledge. He warned, for example, that for respiratory ailments oxymel should not be used because it contained vinegar, which provokes

[6] See chapter 1, p. 31 above, nn. 164 and 165.

[7] *Cartulaire*, document 25. In addition, commentaries on the Hippocratic treatise were written at Montpellier by Magister Cardinalis (fl. 1240), Arnold of Villanova (ca. 1240-1311), and Petrus de Capitestagno (fl. 1299–1330).

[8] "Et iuxta naturam istorum preteritorum sollers medicus conflabit formam regiminis vel considerabit. ... Patet igitur quod prudens medicus ex concursu particularium elicit formam regiminis in omni morbo...": *De regimine acutorum*, in Vatican MS Pal. 1083, fol. 279v (references hereafter are to this MS unless otherwise noted).

[9] [*Incipit*] "Regimen acutarum egritudinum consistit in tribus: in dieta, in pocionibus, et potibus alterativis. ... Primum capitulum de recta administracione cibi et potus in qualitate et quantitate et ordine iuxta arbitrium medici": ibid., fol. 278r.

[10] "Medicinam digestivam [*sic*] quam ponit Ypocras sunt quinque vinum ydromel oxymel aqua et balneum": ibid., fol. 280v.

[11] "Tamen didici a vetulis magis per viam experimenti quam racionis quod malve bismalve pistate cum oleo ... solvunt dolorem in pleuresi": ibid.

coughing, "and in this many have been deceived."[12] The right vein should be struck in phlebotomizing for pleurisy, he asserted, "but in this are mistaken all those who believe that the veins, which nourish the ribs, comprise the vein in the arm: a thing that would not occur to anyone who pays attention to anatomy."[13]

Upon completion of his *Regimen acutorum* Bernard was apparently not satisfied with the quality of his first attempt at writing. He found the treatise "diffuse and lengthy," whereas prolixity is the unkind "stepmother of truth." Therefore he appended, "by way of an epilogue," a *Compendium de regimine acutorum*.[14] This compendium, which he dedicated to his colleague Master Johannes de Confluente,[15] may actually have predated the *Regimen acutorum* itself. According to a colophon in some manuscripts, Bernard had composed the *Compendium* while he was still a bachelor, at the time of his *cursus primus*.[16] In any event, this synopsis of the regimen, like the preceding treatise, is firmly guided by the Hippocratic tradition. Bernard begins his compendium with a list of some errors of physicians. The worst error, he asserts, is to introduce rapid changes in the regimen, since "all sudden change is harmful both to the healthy and to the sick."[17] Hippocrates, too, had stated in his *Regimen of Acute Diseases* that "the greater part of my discourse has related to changes" and their effect on health.[18] Next, Bernard summarizes the

[12] "Summo tamen opere cavendum est quod in passionibus spiritualibus non provocetur tussis. ... Acetosa autem provocant et in hoc sunt multi decepti": ibid., fol. 281r.

[13] "In hoc autem sunt decepti omnes qui credunt quod vene que nutriunt costas constituant venam in brachio quod nusquam reperitur inspicienti anathomiam": ibid., fol. 280r.

[14] [*Incipit*] "Cum omnis prolixitas noverca sit veritatis ideo ego intendens librum regimenti diffuse late quoque conditum sum motus amore mei socii magistri Johannis de Confluente hunc librum a capite breviter sub quodam epilogo pertractare. Regimen acutarum egritudinum in tribus consistit": ibid., fol. 281v. The "*liber*," referred to by Bernard, seems to be his own rather than that of Hippocrates, as Littré would have it, *HL* 25: 323. The opening maxim is cited as "Prolixitas est noverca memorie" in Walther, *Proverbia*, item 22607.

[15] See preceding note. While nothing else is known of Johannes de Confluente, Karl Sudhoff suggested that he may have been of the same Piedmontese family as Jacobinus de Confluentia: "Das dem Magister Johannes de Conflente [*sic*] gewidmete Regimen acutarum aegritudinum des Bernhard von Gordon": *Mitteilungen zur Geschichte der Medizin*, 14 (1915), 93.

[16] "[Ista] fuerunt facta in statu baccalarij tempore cursus primi": *Compendium regiminis acutorum*, in Vatican MS Pal. 1083, fol. 281v.

[17] "Omnis subita mutacio nocet et in sanis et in egris": ibid., fol. 282r.

[18] Hippocrates, *On Regimen in Acute Diseases*, tr. Francis Adams, in *The Theory and Practice of Medicine*, ed. Emerson C. Kelly (New York, 1964), p. 72. The references hereafter to Hippocratic works are to this translation.

correct diet, which consists chiefly of ptisan or barley water, a truly Hippocratic panacea.[19] Then follows a brief description of the digestive beverages and of the purgations by baths and phlebotomy. In one case, Bernard contrasts the advice of Galen, who found phlebotomy safer, with that of Hippocrates, who preferred purgative medicines. Bernard also adjusted Hippocrates' prescription to fit his own climate.[20] He concluded his tract first with the advice that his rules for acute fevers could be adapted to the treatment of chronic diseases, and then with an apologetic plea, "If I have not assembled completely the sayings of Hippocrates, I ask forgiveness for the omissions."[21] Although the *Compendium de regimine acutorum* may have sacrificed completeness to brevity, it nonetheless provided a very practical guide for the physician, and it apparently became quite popular. The author himself, five years after the publication of the *Compendium*, proudly acknowledged that it had "come into the hands of many."[22] With the successful reception of this work, Bernard was launched on a fruitful writing career, which he seems to have combined with his university lectures and medical practice.

Bernard's next work, the *Liber pronosticorum*, was published (*editus*) on 25 January 1295.[23] Although he was probably only in his late thirties at the time, he conceived this book as "a solace for my old age."[24] The apparent contradiction between this statement and his age is resolved,

[19] Ibid., pp. 62-65; in note 3, p. 62, the editor indicates that this treatise was even "quoted by Athenaeus (Deipnos, ii.16) by the name of the work *On the Ptisan*."

[20] "Si materia vadit ad inferiora tunc competit secundum Galenum flebotomia quantum est de securitate, secundum Ypocratem competit farmacia ut elleborus niger repressus cum lauro vel felicinum cum asa et istud dabatur tempore Ypocratis. Et ista sunt perfectius pertractata in lilio [if this is a reference to Bernard's later *Lilium medicine*, it is an interpolation]. Nos vero qui sumus in alio climate non utimur illis medicinis sed ventrem lenimus cum cassiasi et similibus": *Compendium*, fol. 282v.

[21] "Et sic per dei graciam in acute febricitantibus fiet cura. Et tu per viam exempli secundum magis et minus potes idem intelligere in cronicis. Et si complete dicta Ypocratis non agregavi peto veniam de obmissis": ibid., fol. 283r.

[22] "Longa sunt tempora quod nos abreviavimus regimentum acutorum et venit ad manus multorum": *De ingenio sanitatis*, epilogue, in Vatican MS Pal. 1083, fol. 285v.

[23] "Explicit tractatus brevis et utilis supra pronostica editus a magistro Bernardo de Gordonio in preclaro studio Montispessulani anno domini m°cc°lxxxxiiii°. In conversione s. Pauli editus fuit iste tractatus": *Liber pronosticorum*, in Vatican MS Pal. 1174, fol. 103r (references hereafter are to this MS unless otherwise noted). Since the above dating uses the Gallican reckoning, beginning the year with Easter, the date in our calendar is 1295; see Sudhoff, "Schriftstellerei," p. 169.

[24] [*Incipit*] "Senectus domina oblivionis est. Nunc autem cum sciencia predicendi sit ardua et difficilis necnon et multum utilis. Ideo signa pronostica et dies creticos intendo sub quodam epilogo pertractare ut mee sit solacium senectuti non confidens in aliquo de meis viribus sed de ubertate gracie salvatoris": *Liber pronosticorum*, fol. 73v.

however, in the same treatise, since Bernard defined "old age" as starting around thirty-five.[25] The *Liber pronosticorum* or *De crisi et criticis diebus* may have grown out of his lectures, because he stated that he intended to treat the subject matter "by way of an epilogue."[26] In fact, it was based on works of the same titles by Hippocrates and Galen, which were prescribed textbooks in the Montpellier curriculum.[27] However, in his treatise, Bernard alluded to Hippocrates only once, and named Galen twice. He mentioned no other authorities by name. His failure to do so is somewhat surprising in view of the fact that, as he stated in later works, he intended this treatise as a "brief but comprehensive compilation" in which he "faithfully brought together the sayings of Hippocrates, Galen, Avicenna, and others."[28] The most noteworthy feature and improvement on the sources of the *Liber pronosticorum* is the logical organization, by which Bernard brought order and clarity into the "arduous and difficult science of predicting" the course of a disease.[29] In this compilation, as in his preceding *Regimen acutorum*, Bernard was again concerned with brevity. He avoided the detailed treatment of several topics, such as certain fevers and diagnosis by pulse, either because he planned to discuss a particular subject in a later treatise,[30] or simply because he feared the embarrassment of a lengthy book that "would not escape the vituperation of the critic."[31]

[25] "Senectus seu [etas] consistencie durat a xxxv fere usque ad xlx [*sic*]": ibid., 2.6, fol. 83v. Also Hippocrates considered the age of thirty-five as the end of "youth": see his *Prognostics*, ed. Kelly, p. 57.

[26] See n. 24 above.

[27] *Cartulaire*, documents 25 and 68. Commentaries on the *Prognostics* were written at Montpellier by Magister Cardinalis (fl. 1240), Bartholomeus Brugensis (fl. 1307–1330), and Stephanus Arlandi (fl. 1319).

[28] "Postea [viz. after the *Regimen acutorum*] compilavimus opus de crisi et de diebus creticis breviter et cum integritate": *De ingenio sanitatis*, epilogue, fol. 285v.

"Qui autem differencias ... [febrium] scire voluerit ... vadat ad librum nostrum de pronosticis. Ibi enim cum auxilio divino fideliter aggregavimus dicta Ypocratis et Galeni et Avicenne et aliorum": *Lilium medicine*, 1.1, fol. 1r.

[29] See n. 24 above.

[30] The subject of quartan fever "hic autem breviter pertransimus, quia per intencionem cum auxilio Dei alibi intendimus pertractare": *Liber pronosticorum*, 1.3, fol. 76r. On pulses, "Si autem istam materiam pertractaremus particulariter nimis prolongaretur opus nostrum, quoniam alibi cum auxilio divino intendimus pertractare per modum tabularum. ... Excuso me nunc a prolixiori determinacione, quia per Dei graciam largius temptaturus sum hanc arduam materiam aggredi": ibid., 4.5, fols. 90v-91r. Also on the regimen of health, ibid., 2.6, fol. 83r.

[31] On tertian fever, "Et si vellem particularia hec pertractare, nimis prolongaretur opus, et esset dehonestatum cum non evadat liber a vituperacione vituperantis ... nec audeo tamen nimis pertractare timens prolixitatem": ibid., 1.1, fol. 74v.

The result is a work that, while about ten times as long as the *Regimen acutorum*, concisely set down the basic principles of prognostication.

Bernard's first consideration was to determine the nature of the diseases under scrutiny because, "as prognostication precedes treatment, the knowledge [of the disease] precedes prognostication."[32] The diseases to be considered are divided into fevers, compound diseases and abscesses; and the differences between fevers (tertian, daily, and quartan) are unraveled in a separate chapter. The second part of the treatise assesses the prognostic importance of each season, as well as of the location or region and climate, in addition to the patient's age, habits, and "complexion" or constitution. In the third part Bernard defined the nature, length, and intensity of the crisis or paroxysm; while in the fourth part he gave instructions on how to predict the course of a disease from the patient's faculties or *virtutes*. By these *virtutes*, to each of which he devoted a chapter, Bernard meant the mental faculties (thinking process, imagination, and memory), as well as the five senses, appearance and behavior (including movement and sleep), and body functions (breathing, pulse, excretions, and so on). The fifth and final part of the treatise relates to the nature and prediction of crises and critical days, that is days on which the crisis was likely to occur.

This last portion of the *Liber pronosticorum* is of particular interest because it centers, especially in the latter half, on medical astronomy and astrology.[33] Within the last four chapters of this section, Bernard determined the critical days on the basis of the lunar calendar, and described the "nature and complexion of the heavenly bodies insofar as these pertain to the physician." He also gave the rules for computing the "medical week and month." In the final chapter he explained the difference between the "solar month" and the "lunar month": the former, based on a convenient division of the solar year, is followed in the "vulgar" or ecclesiastical computation; the latter, however, is based on the moon's course and should be observed by the physician with regard to critical days. These chapters, in which Bernard lucidly summarized the theory and practice of medical astrology, could serve as convenient guides

[32] "Sicut pronosticacio antecedit operacionem, ita cognicio pronosticacionem": ibid., 1.2, fol. 75r. The same idea is expressed in the *incipit* of the text of the treatise, "In morbis enim pronosticare non possumus nisi morbum cognoscamus": ibid., fol.74r.

[33] The special interest of these chapters was already noted by the scribe of an early manuscript: "Explicit tractatus super pronostica Ypocratis et de crisi et creticis diebus Gallieni et continens aliquid de astrologia que est necessaria medico...," in Munich Codex Latinus Monacensis (Clm) 26908, first half 14th c., fol. 91, cited by Sudhoff, "Schriftstellerei," p. 169, n. 2.

to the physician who was less well versed in the subject. Their usefulness in this regard may explain why they seem to occur as individual tracts in several manuscripts.[34] One of these, the chapter "*De natura et complexione superiorum*" (5.7) deserves special mention because Charles Singer, while attributing it to Guy de Chauliac (ca. 1290–1368), has given a pertinent evaluation of its form and content. Dr. Singer found the style "rather heavy and labored" and the classification in parts, sections and chapters "excessive." He considered the contents "entirely typical of the Arabian school," but he also remarked that the text "has a quality that is rare in the writings of that school: it is quite brief and does not display the 'verbositas arabica'."[35] With regard to Arabian influences, it is worth noting that also the Hippocratic writings had earlier considered the connection between the heavens and the critical days,[36] even though Bernard appeared to have drawn his materials from pseudo-Hippocratic and pseudo-Ptolemaic or Hermetic sources; he may also have borrowed from the pseudo-Aristotelian *Secretum secretorum* or, perhaps, from Albertus Magnus.[37] In any event, his treatment is remarkable not only for his sober concern with medical prognostication but also for his omission of the horoscopic techniques and magical trappings that characterized Hermetic and pseudo-Ptolemaic astrology.[38]

In view of Bernard's professionalism in his *Liber pronosticorum* one is surprised to find that, shortly after the publication of that book, he occupied himself with a purely horoscopic work that, in addition, contains little medical material. On 22 December 1295, in the thirteenth year of his teaching, he translated the *Tractatus de reductione geomancie*

[34] Thus, 5.6, *incipit* "Numerus et sufficiencia dierum creticorum habetur...," in Prague MS 243 (I.F.11), s.d., fol. 176r (TK 960); 5.7, *incipit* "Circa istam materiam est intelligendum quod natura...," in Bristol City Reference Library MS, a. 1420-1435, fol. 272v ; and 5.8, *incipit* "Medicus autem velit nolit supponendo aut quoquomodo fit...," in Cambridge MS University Trinity College 1122 (O.II.18), 15th c., fol. 159v (TK 863).

[35] Charles Singer, "Sur un manuscrit attribué à Guy de Chauliac," *Comptes rendus du deuxième Congrès international d'histoire de la médecine* (Évreux, 1922), pp. 168-169. In the Bristol MS (see preceding note), discussed by Singer, Bernard's chapter (fols. 272v-275v) is indeed preceded by Chauliac's *Magna chirurgia*, but it is followed (fols. 276r-279v) by another excerpt (2.1-4) from Bernard's *Liber pronosticorum*.

[36] See especially Karl Sudhoff, "Zur Geschichte der Lehre von der kritischen Tagen im Krankheitsverlaufe," *Archiv für Geschichte der Medizin*, 21 (1929), 1-22.

[37] This is a possibility, which dawned upon me while reading Thorndike, *Magic*, 2: 587, and while looking over a few dubious "Aristotelian" references in Bernard's writings. For numerous references to MSS of Hippocratic, pseudo-Hippocratic, and Hermetic treatises on astrological medicine, see TK index, 1820-1823.

[38] Such as necromancy, the casting of images, and geomancy; see Thorndike, *Magic*, 2, chapter 45 and passim.

ad orbem. There can be little doubt that Bernard was the author of this translation, because it is dated with unusual precision and accuracy, with the inclusion of the reigning potentates: Pope Boniface [VIII, 1294–1303], Emperor Adolf [of Nassau, 1292–1298], King Philip [IV, 1285–1314] of France, King Jayme [II, 1276–1311] of Mallorca, Bishop Berengar [1290–1298] of Maguelone, and King Edward [I, 1272–1307] of England.[39] In the opening phrases this treatise is heralded as "the secret of the great God revealed to Ptolemy, king of the Arabs," and as a device "to know accurately the present, the past and the future."[40] We may assume that Bernard made this Latin translation from the Provençal,[41] since there is no evidence that he was familiar with any other language. The original text, however, appears to have been in Castilian, since the initial model *interrogatio* refers to the king of Castile: "the question is about the name, the life, the state, the reign, and the overall condition of the king of Castile until the end of his life"[42] – even though King Sancho IV (1284–1296) is not listed among the reigning monarchs in the colophon. In any event, the fact that Bernard took the time to latinize this vernacular work seems to indicate that he wanted to make it more widely known or more easily accessible to scholars, although at least one scholar among his colleagues, Ramon Lull, was to condemn the art of geomancy and its practitioners two years later, in 1297.[43] The *Tractatus de reductione geomancie* had less

[39] "Translacio magistri Bernardi de Gordonio phisici; regnante bonifacio papa et adulpho electo in imperatorem et philippo rege francie filio quondam philippi regis francie, et iacobo rege maioricarum, et berengario episcopo magalonensi, et audoardo rege anglorum; et anno xiii° lecture nostre in phisica in montepessulano, et anno domini m cc nonagesimo quinto, die jovis ante natale," in Paris MS BN lat. 15353, 13-14th c., fol. 92v. The early date of this MS is an additional argument for Bernard's authorship of the translation.

[40] "Incipit archanum magni dei revelatum tholomeo regi arabum de reductione geomancie ad orbem. [P]er hoc presens opus possunt cognosci exquisite presencia, preterita et futura...," *De reductione geomancie ad orbem*, in Paris MS BN lat. 15353, fol. 87r.

[41] A Provençal MS in Paris, BN fr. 14771, ca. 1300, contains divinations whose contents and procedures are very similar to those found in Bernard's translation. In addition, a somewhat later Paris MS, BN lat. 7420A, probably written at Montpellier, contains several Provençal poems on geomancy which are almost certainly inspired by a work similar to the *Reductio geomancie*; the same MS also contains a Provençal tract on astrology, with illustrations strikingly similar to those which adorn Bernard's translation. The two MSS are described by Paul Meyer, "Traités en vers Provençaux sur l'astrologie et la géomancie," *Romania*, 26 (1897), 225-275.

[42] "Queritur de nomine et vita et statu et regno et tota disposicione regis castelle usque ad finem vite sue," *Reductio geomancie*, fol. 87r.

[43] *Tractatus novus de astronomia*: see Thorndike, *Magic*, 2: 869, and Sarton, *Introduction*, 2: 909.

scholarly than practical value, however, as it was designed for calculating future or hidden events such as the weather, the outcome of a voyage, or the whereabouts of stolen property. This treatise almost falls into the category of "ready-reckoners," which common practitioners often wore from their belts.[44]

In the introduction to the allegedly Ptolemaic treatise, the author explains the complex *modus operandi* of the "sacred science" of geomancy, giving instructions on how to draw the "houses," how to assign a heavenly body to each, and how to project the fortunes.[45] He also warns that any questions to be answered must be about "a licit, honest, and carefully weighed matter" and without the intention to deceive.[46] Moreover, the author continues, since the stars are governed by God, the consultation ought to begin with this humble prayer:

> May the grace of the Holy Spirit assist us and show us the Lord's divine judgments; and also: God, who to your blessed Apostle Paul opened the four secrets of heaven and ordered him to reveal them to no one, make me thus solve this question that the devil may have no power to harm me, and send from heaven your holy angel who released Peter from his bonds, that he may guard me and teach me to solve this question through our Lord Christ. Amen.[47]

This prayer formula reveals the Christian and somewhat gnostic origin of the treatise. The first half of the work is copiously illustrated with tables and charts. It also contains definitions of the properties of the elements used in the calculations. Thus, each of the twelve "houses" through which the heavenly bodies pass, is given its own sphere of influence over certain

[44] Compare C. H. Talbot, *Medicine in Medieval England*, p. 126.

[45] "Modus autem operandi in hac sacra sciencia est huiusmodi: fabricantur in primis x domus et in qualibet domo sint iv linee punctorum, et puncta sint proiecta a casu et fortuna sicut fit in geomancia, et prima domus attribuatur geomancie, secunda soli...": *Reductio geomancie*, fol. 87r. While this art of "reduction" appears different from geomancy proper, it would lead us too far afield to attempt to unravel here its complexities. A helpful description and bibliography on geomancy is found in Thorndike, *Magic*, especially 2: 110-121 and 835-837.

[46] "Cum ergo volueris facere questionem oportet quod sit de re licita et honesta et diu cogitata et per intencionem et non ad illudendum": *Reductio geomancie*, fol. 87r.

[47] "Quia procedunt omnia a summo culmine rerum quo non consulto nichil est quod sidera possint ideo in principio dicatur humiliter talis oracio: Sancti Spiritus assit nobis gracia ut ostendat nobis divina Domini iudicia; et cum hoc ista: Deus qui beato Paulo apostolo tuo iv aperuisti secreta celi et precepisti ut nemini revelaret sic me facias solvere questionem istam ut diabolus non habeat potestatem nocendi sed mitte angelum tuum sanctum de celo qui solvit Petrum a ligaminibus, qui me custodiat et doceat solvere questionem istam per Christum dominum nostrum amen": ibid.

groups of events, for example, birth and travel (first house), gain and loss (second house), quarrels and wars (seventh house), and promotion and demotion (tenth house). Next, each of the twelve constellations and each of the nine "erratic planets" is described as to its various qualities: complexion, that of the moon, for example, being "cold and humid"; gender, with Taurus as well as Capricornus characterized as *femineus*; and geomantic symbol, consisting of a geometric configuration of points that has no relation to the more familiar zodiacal figures.[48] In addition, the terrestrial area and period of life influenced by each constellation is given; and each planet is assigned a triple term of effectiveness.[49]

The second half of the treatise, which contains the solutions for over forty particular questions, differs markedly from the first part. The answers to the questions are determined by the conventional astrologers' art, according to the positions of the planets and signs in each zodiacal house, without any reference to geomantic symbols or calculations. Whatever may be the explanation of this discontinuity of the two parts, another feature of the work is more directly relevant to the present study. Some of the inquiries in the second part of the *Reductio geomancie* pertain, directly or indirectly, to medicine. Within the latter category are such questions as, whether a woman can conceive, whether she is pregnant, whether the child will be a boy or a girl, and whether the newborn infant will live. Two other questions are, however, of more vital concern to the physician, namely whether the patient will die and, particularly, the question "whether or not it is good for the physician to agree to treat a patient or even to go and see him."[50] Since physicians usually dreaded the chance of being called for a hopeless case, those two questions may have stirred Bernard's interest in the *Reductio geomancie*.

After translating the above pseudo-Ptolemaic horoscope, Bernard apparently waited for almost four years before producing another work. In July 1299,[51] "in the distinguished *Studium* of Montpellier," he

[48] Thus Aquarius is symbolized by ˙.˙, Leo by ⋮, the sun by ˙.˙ .˙., the moon by ⋮ ⁞⁞, and so on.

[49] For Saturn, for example, "primus terminus xxx annos, secundus xliii et dimidium, ultimus lviij": ibid., fol. 88v. Professor Lemay has pointed out that the lengths of these terms agree with those found in Albumasar's *De annis fardarat planetarum*, tr. John of Seville, but that they are designated there as *minus*, *medius* and *maior*.

[50] "Cap. secundum VI domus. Si vis scire utrum sit bonum medico vel non accipere pacientem in cura sua vel ire ad eum ... si boni planete inveniantur in domo VII [*sic*] bonum est et tunc dogmata Ypocratis verificabuntur ad unguem": *Reductio geomancie*, fol. 91v.

[51] Rather than in 1296, as in the incunabulum edition (Venice, 1498, Hain 7800), fol. 2v, and in the *HL*, 25: 321.

completed his "arrangement" of the methods of treatment and cure of diseases, *De decem ingeniis curandorum morborum*, probably at the request of some of his colleagues.[52] The materials for this treatise were, according to the statement in the manuscripts, "extracted from the sayings of Galen in the *De ingenio* and especially from the final chapter of the twelfth *particula* of that work, and from the first chapter of the fourth *fen* of the first book of Avicenna's *Canon*."[53] This very detailed acknowledgement, in addition to the brevity of the work itself, might tempt one to presume that it was merely a summary of two specific chapters in the works of Galen and Avicenna. However, a closer inspection of the contents reveals references to at least five cited works of Galen,[54] as well as several statements drawn from Hippocrates and Haly Abbas. To these, moreover, Bernard added several of his own ideas. The latter comprised the necessary qualifications of the physician and his assistants, together with a few personal observations that strikingly illustrate Bernard's keen powers of perception and his knowledge of the local area. For example, when he repeated Galen's prescription of aromatics for the purgation of the brain Bernard recommended "the peach, when not yet red, especially the kind found in Montpellier."[55] Also, with regard to a healthy environment he advised the physician to "check whether the patient was disturbed by an adjoining shop occupied by a carpenter, hammerer, tanner of hides, melter of tallow, or by persons who work with sulfur and the like."[56] Thus, Bernard adapted to his immediate environment the doctrines of Galen's *De ingenio sanitatis*, a work that held a prominent

[52] "Facta fuit hec ordinacio in preclaro studio montis pessulani anno domini m°cc° nonagesimo nono mense iulii die mercurii post festum sancti marcialis, et quia aliqui ex sociis volunt scire quod dandum est...": *De decem ingeniis curandorum morborum*, in Munich Clm 340, a. 1304, fol. 264v, quoted by Sudhoff, "Schriftstellerei," p. 167. It may be noted that the reference to the "*sociis*" seems somewhat ambiguous; see the variant in the following note.

[53] "Et ista omnia extracta sunt ex dictis Galeni de ingenio et potissime ex xij particula et potissime ex capitulo finali, et ex dictis Avicenne primo canonis fen quarta capitulo primo, et quia aliqui ex sociis volunt scire quid dicendum est...": *De ingenio sanitatis*, in Vatican MS Pal. 1083, a. 1447-1458, fol. 285v.

[54] Besides Galen's *De ingenio*, Barnard also cited his *Tegni* and his commentaries on the Hippocratic *Regimen acutorum*, *Prognostications* and *Aphorisms*.

[55] "[Ad cerebrum confortandum] debemus que sunt boni odoris exhibere et pro tanto valet persecum rubeum novum est potissime inventum in monte pessulano": *De ingenio sanitatis*, in Vatican MS Pal. 1083, fol. 284v. The 1498 incunabulum has "*poma rubea*" instead of "*persicum*," fol. 2r.

[56] "Et debet advertere medicus si est officium iuxta pacientem contristans ipsum sicut sunt fabri et malleatores et paratores coriorum et fundentes sepum, et qui operantur de sulphure et similibus": *De ingenio sanitatis*, in Vatican MS Pal. 1083, fol. 283r.

place in the Montpellier curriculum for it was listed both in the 1309 syllabus for students and in the 1340 course program for masters.[57]

Undoubtedly because he wanted his works to be useful to his students and colleagues, one of Bernard's more striking contributions in the treatise *De decem ingeniis*, as in his earlier writings, is his arrangement or organization of the subject matter. He himself later referred to this work as a "table of methods."[58] He divided the treatment of diseases into ten methods (*ingenia*) based on different "approaches" (*consideraciones*) or aspects, such as, for example, a knowledge of the specific disease and its medication, and attention to the patient's "complexion." Another method of treatment was governed by the consideration of the organs and their qualities, namely strength, composition, location, and sensitivity. Bernard further subdivided these principal categories or methods and placed each of them under particular rules, as a systematic guide for the diagnosing physician.[59] This rigid arrangement, apparently intended as a handy resume of medical theory as well as for a diagnostic checklist, had one drawback. In his endeavor to distinguish between categories and their subdivisions, Bernard seems to have been so carried away by subtle logic and conceptualization, as to make his treatise resemble an exercise in scholastic sophistry. Nevertheless, the *De decem ingeniis* is interesting for its combination of deductive reasoning and practical application.

A similar combination of practical concern and theoretical approach characterizes the *Tractatus de gradibus*, which Bernard "arranged" in the first half of 1303.[60] This work deals with the physician's practical problem of determining the probable effect of each specific drug that he wished to administer. The normally knowledgeable physician, even when applying the traditional principle that "contraries cure," and classifying each drug according to its primary qualities (hot or cold, dry or moist) and its specific

[57] *Cartulaire*, documents 25 and 68.

[58] "[Addatur] libellus qui intitulatur tabula ingeniorum": *Lilium medicine*, epilogue, fol. 99v; also see n. 88 below.

[59] [*Incipit*] "Ingenium (Ingenia: Paris BN 16189 etc.) curacionis morborum quantum est de presenti sunt x. Primum sumitur ex morbo. ... Prima consideracio ex morbo sumpta fortis et valida. ... Ingenia curandi morbos sumpta ex morbi essencia sunt xi. Primum sumitur ex morbi essencia scilicet si est simplex vel compositus ... si simplex, an intensus an remissus. Et si remissus debile medicamen sufficit...": *De ingenio sanitatis*, fol. 283r.

[60] "Incipit tractatus de gradibus ordinatus a magistro Bernardo de Gordonio decano medicorum in almo studio montis pessulani a.D. M°ccc°iij°," in Erfurt MS Amplonianum F.236, first half 14th c., fol. 50v, cited by Sudhoff, "Schriftstellerei," p. 179. We may infer that the work was finished in the first half of 1303, since Bernard began the *Lilium medicine* in July 1303.

properties (for example, constrictive or laxative),[61] needed, nonetheless, to know the intensity or degree of the drug's effectiveness. Until the thirteenth century this problem had been approached empirically, but Bernard now elaborated a "precise and detailed doctrine" of gradation.[62] He realized that, while the intensity of simples could be learned from experience, "that process would be too lengthy" for compound medicines, "and meanwhile the patient would be put in great jeopardy." Hence, Bernard held, it was "necessary to have a rule of thumb and a guide which the physician could readily utilize for compounds." Such a standard could be found "only by reason."[63]

The standard or rational guide, developed by Bernard in his *Tractatus de gradibus*, consisted of a mathematical theory, namely the quantification of medicinal qualities.[64] Bernard did not claim to have invented this theory but admitted that, so far as he knew, the science of gradation was founded by four "illustrious men," namely Galen, Jacob Alkindi, Averroës, and Master Arnold of Villanova. Appropriating a phrase from Saint Paul, Bernard humbly admitted that, although unworthy, he had been given the honor of "collecting and, in a way, disclosing the secrets" of this science.[65] Only a few years earlier Arnold of Villanova had indeed

[61] These qualities and properties could be found in alphabetical lists of simples, such as the anonymous but popular *Liber graduum*; see Loren MacKinney, "Medieval Medical Dictionaries and Glossaries," in *Medieval and Historiographical Essays in Honor of J. W. Thompson* (Chicago, 1938), pp. 259-260.

[62] After describing a prescription to control flegm, Bernard added, "ista autem graduacio est satis grossa, et hec sufficit propter iuvenes, quoniam exquisite et in puncto tradita est doctrina in libello de graduacione ... brevior tamen et facilior modus graduandi propter iuvenes est iste...": *De conservacione*, 2: *De urinis*, 8, in Vatican MS Pal. 1174, fol. 35v; also, "recurramus ad libellum nostrum de gradibus, ubi perfecte tradita est natura graduacionis": ibid., fol. 35v.

[63] "Per viam autem sensus potissime simplicia cognoscantur. Si autem composita per viam experimenti vellemus cognoscere, via esset nimis longa et paciens interim in magno periculo poneretur. Oportet igitur quod sit aliquid medicum regulans et dirigens, quomodo possit medicus breviter cum compositis operari; hoc non potest esse nisi cum ratione...": *De gradibus*, ed. Julius Pagel, "Über die Grade der Arzneien nach einer bisher ungedruckten Schrift des Bernhard von Gordon aus dem Jahre 1303," *Pharmaceutische Post*, 28 (1895), 182. Further references are to this edition, unless otherwise noted.

[64] The history and implications of this theory are discussed fully in a scholarly article by Michael McVaugh, "Quantified Medical Theory and Practice at Fourteenth-Century Montpellier," *Bulletin of the History of Medicine*, 43 (1969), 397-413.

[65] "Illustrissimi viri qui hanc scientiam fundaverunt, quorum notitia pervenit ad nos, fuerunt quattuor: primus fuit Galenus, secundus fuit Jacobus Alkindi, tertius fuit Averroes, quartus fuit magister Arnoldus de Villanova. Mihi autem tamquam abortivo data est gratia [1 Cor. 15.8] colligendi et secreta aliquo modo reserandi...": *De gradibus*, pp. 143-144.

formulated a theory of gradation in his *Aphorismi de gradibus*.[66] While Bernard deferred to his colleague as a founder of the science, his own treatise was, according to Professor McVaugh, "intended to bring the abstruse *Aphorismi* down to a less technical level of general application."[67] In addition, Bernard provided answers for the scientific objections that might have been raised against Arnold's system, particularly the objection that a solid experimental foundation could never be established for a theory of gradation. He strongly questioned the alleged conflict between reason and sense-experience. With emphatic irony, or perhaps with genuine emotion, "for love's sake," he asked how a rational doctrine of gradation could possibly be formulated *without* an empirical basis, "*nisi solum per experimentum*."[68] Bernard realized fully, however, that the dilemma between an empirical and a rational approach was not so easily solved. Therefore he described the first part of his *Tractatus* as "full of bitter anguish for the intellect."[69]

The second part or *doctrina* of the treatise *De gradibus* was to be a "comfort and delight" for the mind, since it proceeded by the empirical method, "*per viam experimenti*."[70] It was avowedly "based on Galenic roots," and listed five general conditions and five specific rules to be observed in conducting an experiment. Thus, for example, the degree of hotness and dryness of pepper cannot be measured when the pepper is frozen: the ingredient to be tested should be "free of any accidentally acquired property"; otherwise "the experiment would be fallacious."[71] The third and last part of the treatise "will be of great usefulness and thus satisfy the intellect," Bernard promises. In this part he describes, in

[66] MSS in TK 691. Similar treatises were written at Montpellier by Stephanus Arlandi (ca. 1319), Jordanus de Turre (ca. 1330), and Gerardus de Solo (1371).

[67] McVaugh, "Quantified Medical Theory," p. 402.

[68] "Quam amara igitur erit natura graduationis, cum intellectus in eam non valeat penetrare! Non est ergo via nec doctrina aliqua quae possit nos ducere in naturam graduationis nisi solum experimentum et ita videtur deperire doctrina omnium de natura graduationis pertractantium. ... per amorem nunc ergo quaero: quis intellectus, quae ratio in naturam horum pervenire potest per viam rationis certe non nisi solum per experimentum?" *De gradibus*, pp. 131-133.

[69] [*Incipit*] "Doctrinam de gradibus intendimus tripliciter ordinare, quarum prima est tristabilis cum amaritudine intellectus, secunda erit magnae consolationis, in qua delectabitur intellectus, tertia erit magnae utilitatis, in qua satiabitur intellectus": ibid., p. 131.

[70] "Secunda doctrina est quae procedit per viam experimenti et fundatur supra radices Galeni in fine de complexionibus et in fine v. particulae de simplici medicina": ibid., p. 143. Also see n. 68 above.

[71] "Oportet v conditiones observare; prima est haec, quod res sit denudata ab omni qualitate accidentaliter acquisita ... aliter exprimentum erit fallax": *De gradibus*, p. 143.

accordance with the above mentioned "founders," the detailed doctrine of gradation and the calculatory techniques to be used. Where pertinent, he also gives concrete examples, such as that of various wines whose degree of potency or "calefacient" effect depends on their region of origin.[72] The *Tractatus de gradibus*, in addition to his earlier treatises, provides substance for the view that Bernard contributed much of his own thought. In fact, he himself asserted that it is the mark of "a very poor mind" simply to copy what one found.[73] Therefore, he should perhaps not be taken too literally when he modestly calls his own works "compilations" of the sayings of other authors. The same observation may be applied especially to his compendium, the *Practica seu Lilium medicine*.

The *Lilium medicine* was begun in the *Studium* of Montpellier in July 1303, after Bernard's twentieth year of teaching,[74] and completed (*scriptum*) twenty months later, on 5 February 1305.[75] It may be considered his *magnum opus* because it was not only the longest and most comprehensive of his works but also the most famous. The *Lilium*'s fame is manifest in its fate during the succeeding centuries: the Latin text is extant in more than fifty manuscripts and six printed editions, and it was translated into French, German, Hebrew, and probably Provençal during the fourteenth century and into Spanish and Irish during the fifteenth century.[76] The work was also listed as required reading for medical students at the University of Montpellier shortly after 1400,[77] and at the

[72] Ibid., p. 260.

[73] "Miserrimi ingenii est uti semper inventis": *De gradibus*, p. 144.

[74] "Inchoatus autem est liber iste cum auxilio magni dei in preclaro studio montis pessulani post annum xx. lecture nostre, anno domini m°ccc°iij° mense julii": *Lilium medicine*, prologue, fol. 1r. I am preparing a critical and annotated edition and an English translation of the *Lilium*.

[75] "Fuit scriptus iste liber anno incarnacionis domini [beginning 25 March] m°ccc°iiij die sancte agathe virginis": ibid., fol. 99v. Since this colophon occurs in several other MSS, *scriptus* evidently refers to the composition rather than to the copying. Hence it is misleading to date those MSS as AD 1305, as is done in several catalogues and by Rolf Winau, "Eine Lilium medicinae-Handschrift aus dem Besitz des Nikolaus von Kues," *Medizin-historisches Journal*, 1 (1966), 28 (with AD 1304 as an additional misreading). The above colophon, which confirms the date 1305 given in most printed editions of the *Lilium*, was unfortunately unknown to Dr. Sudhoff when he wrote on Bernard's "Schriftstellerei" (pp. 163-165) in wartime Germany; it was perhaps overlooked by Ynez O'Neill for her article, "Dates in the Printed Editions of the Lilium Medicinae," *Sudhoffs Archiv*, 49 (1965), 86-89. Possibly the oldest extant MS is BN lat. 6964, which was "made" on 14 November 1305 according to the colophon, "Factum anno m°ccc° quinto die lune post festum beati brixii" (fol. 81v).

[76] See appendix below.

[77] See Ernest Wickersheimer, "Le discours de réception d'un bachelier en médecine

University of Vienna in 1520.[78] In addition, the manuscript was a prized possession in some private libraries, such as that of the author's more famous colleague, Arnold of Villanova.[79] Cardinal Nicolaus Cusanus (1401–1464) owned a copy in which he inscribed marginal notes,[80] and the fifteenth-century Malines physician Arthur van Malderen even had two copies in his personal library.[81] Because the *Lilium* was so celebrated, the conspicuous absence both of a modern printed edition and of an English translation is ironic, and I hope to remedy this lack in the near future.

Bernard himself seems not to have aimed for such glory as the *Lilium medicine* was to reap for generations. He declared in his prologue that he intended only "to consider some easy, common, and useful matters for the benefit of the humble, in other words to compile a book about medical practice."[82] He emphasized that he was "not ashamed to repeat for the humble what is commonly said about practice."[83] Three years later he happily recalled that "it was not long ago that I wrote the *Lilium medicine* for the humble," and it was received favorably.[84] The emphasis on

---

montpelliérain au début du xve siècle," *Bulletin de la Société française d'Histoire de la Médecine*, 13 (1910), 247. The bachelor in question was Heinrich Lamme of Lübeck.

[78] See Richard Durling, "An Early Manual for the Medical Student and the Newly-fledged Practitioner: Martin Stainpeis' *Liber de Modo studendi seu legendi in medicina*," *Clio Medica*, 5 (1970), 19.

[79] See Michael McVaugh, "Nota sobre las relaciones entre dos maestros de Montpellier: Arnau de Vilanova y Bernardo Gordon," *Asclepio*, 25 (1973), 332.

[80] London MS BL Harl. 3698, fol. 99v contains a note, "N de Cusa qui emit a magistro fabricio de Lorch istum et alios plures libros de medicina et artibus, qui fuerunt illius magistri petri rodmullir plebanus ibidem, quod 50 floribus renanibus. 1449."

[81] See A. Louis, "Analyse des bibliothèques de deux médecins malinois du xve siècle," *Janus*, 53 (1966), 241-305. The second physician, Canon Simon van der Sluys, also possessed a copy of the *Lilium*.

[82] [*Incipit*] "Interrogatus a quodam Socrates quomodo posset optime dicere respondit si nichil dixeris nisi quod optime sciveris. Nichil autem scimus nisi quod a nobis frequenter dictum est et quod commune est ... ideo confidens de domino scienciarum quedam facilia et communia et utilia intendo ad utilitatem humilium pertractare, librum scilicet de practica compilare": *Lilium*, prologue, fol. 1r.

[83] "Et quia de humilibus (humilibus Paris BN 6964, fol. 1r) scribo, ex hoc superbi abiecti sunt. ... Cum a nobis (cum aliis BN 6964) scripta communia contempnunt ymmo erubescunt dicere aliquid ex quo semel dictum est. ... Quia igitur memoria hominis labilis, quod communiter de practica dicitur repetere humilibus non erubesco": ibid.

[84] "Reduco ad memoriam quod non est multum temporis quod scripsi humilibus lilium medicine, et quia sine livore et cum benignitate receptus est, ideo hoc opus aggredi libentius attemptabo": *Liber de conservacione vite humane*, prologue, in Vatican MS Pal. 1174, fol. 10r. The phrase *scripsi humilibus* was incorrectly translated as "I wrote in all humbleness," by H. I. Bell in "The Bibliography of Bernard de Gordon's 'De conservatione vitae humanae'," *17th International Congress of Medicine* (London, 1913), p. 328.

humility was not merely the rhetorical *captatio benevolentiae* of a classical prologue, but apparently also a very personal feeling of Bernard's and, in addition, an echo of the Hippocratic tradition. For example, in the treatise *On Ancient Medicine*, attributed to Hippocrates, there is the assertion that "whoever treats of this art should treat of things which are familiar to the common people," and that "whoever does not reach the capacity of the illiterate vulgar and fails to make them listen to him, misses his mark."[85] Moreover, Bernard evidently destined his *Lilium* for the young and inexperienced physician, for he went into basic details of pathology and therapeutics expressly for the sake of "young colleagues" even though, as he observed, "the advanced do not need" such details.[86] At the same time he omitted treatments that could only be handled by the "competent and very expert physician,"[87] and he referred the advanced reader to his more specialized works for several topics.[88]

Hence, in view of Bernard's stated purpose and procedure, it seems unfair to charge that the *Lilium medicine* "lacks depth and thoroughness,"[89] or that it "does not distinguish itself anywhere by anything new."[90] On the contrary, while one may be skeptical as to the originality of alleged "firsts" in this compilation, one may also be appreciative of their appearance in this context. For example, some recent historians have credited the *Lilium* with the first mention of eyeglasses in a medical textbook.[91] Others have praised Bernard's originality in his application of a

[85] *On Ancient Medicine*, tr. Adams, in *Theory and Practice*, ed. Kelly, p. 2.

[86] Thus, against consumption, Bernard gives prescriptions which "sufficerent quantum ad *provectos*; propter tamen *iuvenes* magis particulariter procedamus": *Lilium*, 4.5, fol. 52v. Also, "quia *socii iuvenes* volunt quod fiat distinctio ... ideo volo magis particulariter pertractare et ordinare, licet *provecti* non indigeant": ibid., 4.7, fol. 54v (italics mine).

[87] After giving a complex remedy for wounds, Bernard adds, "sed in hoc oportet quod medicus competens esset et multum expertus. Ad alia igitur currendum est": ibid., 1.28, fol. 18r.

[88] For example, the diagnostics of pain "satis declaravimus in tabula ingeniorum," and its treatment "oportet se fundare in istis radicibus sicut manifestabitur cum de regimine sanitatis scribemus": ibid., 1.17, fol. 9r. On diarrhea, "qui autem plus de ista materia scire voluerit, vadat ad librum nostrum de pronosticacione": ibid., 5.16, fol. 72r. And so on, passim.

[89] William Lennox, "Bernard of Gordon on Epilepsy," *Annals of Medical History*, 3 (1941), 372.

[90] "... cette pratique ne se distingue en aucun endroit par quelque nouveauté, soit dans l'arrangement des matières, soit dans les aperçus d'un esprit dont l'originalité a sa source dans la méditation prolongée ou dans l'expérience acquise": Littré, *HL*, 25: 326.

[91] Among numerous others, Sarton, *Introduction*, 3: 874. This assumption is challenged in a detailed study on the subject by Edward Rosen, "The Invention of Eyeglasses," *Journal of the History of Medicine and Allied Sciences*, 11 (1956), 201-202. Bernard prescribed an eyedrop which "est tante virtutis quod decrepitum faceret legere

truss for hernia,[92] and in his diagnosis of epilepsy in general and of *petit mal* in particular.[93] However, whether or not these were original with him is uncertain. The originality of the *Lilium*, as well as the reason for its popularity, would seem to lie in its lucid conciseness and in the arrangement of its subject matter. Bernard frequently reminded his reader that "we are concerned with brevity."[94] Desirous of keeping his work brief as well as practical, he avoided speculation on several questions.[95] Furthermore, he seemed aware of the fact that it would be futile to attempt complete coverage of the individual symptoms, circumstances, and prescriptions in given cases, since "according to Galen not everything can be denoted by specific terms."[96] Bernard, therefore, asserted that, if anything had been omitted, he would leave it up to "the kindness and diligence of the reader to supply it."[97] Finally one might suggest that it seems somewhat unkind to blame the *Lilium* for superficiality and "fondness for 'shot-gun' prescriptions,"[98] when Bernard explicitly stated that he had left the filling in of particulars "up to the conscience of the practising physician."[99] Moreover, he continued, since he offered here only a "doctrine of practice," the reader who was "endowed with a head" should be wise and judicious enough to adjust the quantity and quality of

literas minutas sine oculo berellino": *Lilium*, 3.5, fol. 39v-40r. In the Venice incunabulum of 1498 (fol. 41v), as in most early printed versions, *oculo berellino* has been replaced by *ocularibus*.

[92] Talbot, *Medicine in Medieval England*, p. 104. However, Talbot seems to contradict himself when he mentions a truss, prescribed by William de Congenis (early 13th c.), p. 94; and see also pp. 89-90. Bernard recommends "bracale ferreum cum lingula ad modum semicirculi": *Lilium*, 7.7, fol. 91v.

[93] Lennox, "Bernard of Gordon on Epilepsy," pp. 372-383. Less appreciative is the evaluation by E. Pirkner, "Epilepsy in the Light of History," *Annals of Medical History*, 1 (1939), 453-480. Bernard's diagnosis of *petit mal* which, as he claims, he had observed frequently, is found in the *Lilium*, 2.25, fol. 33r. His observations and prescriptions are cited in their historical context by Owsei Temkin, *The Falling Sickness*, 2nd ed. (Baltimore, 1971), pp. 121-122 and passim.

[94] There are so many kinds of hemorrhage of the liver that "ad ea narrandum non sufficiet presens dies, sed ista sufficiant causa brevitatis, quia circa brevitatem laboramus ... et propter brevitatem potest reduci ad duas causas": *Lilium*, 6.4, fol. 79v.

[95] For example, on the question whether coitus cures daily fever: "de ista autem materia nolo amplius raciocinari, timens prolixitatem": ibid., 1.2, fol. 2r; also, "speculacionibus pretermissis": ibid., 4.9, fol. 57v; and "nunc non est locus raciocinandi": ibid., 5.2, fol. 61r.

[96] "Secundum Galenum omnia certis literis denotari non possunt": ibid., 4.5, fol. 52v; also passim.

[97] "Et si aliqua obmisi, suppleat hoc diligens amicicia lectoris": ibid., 4.4, fol. 51r.

[98] Lennox, "Bernard of Gordon on Epilepsy," p. 261.

[99] On the cures for tertian fever, "opiniones tamen sunt de ista materia, et ideo consciencie operantis derelinquo": *Lilium*, 1.4, fol. 3r.

prescriptions to each individual case.[100] In one instance omitting pharmaceutics in the cure for enuresis Bernard even admitted that he had, "for the sake of brevity, not kept the usual arrangement."[101]

As a rule, however, Bernard was religiously consistent in arranging his subject matter because he believed that "it would not be proper (*decens*) to put anything just anywhere."[102] In addition he constantly gave cross-references to pertinent chapters or passages. The insistence upon a logical organization may indeed be considered his foremost contribution in the *Lilium*. The work is divided into seven *particulae* of which the first, dealing with fevers, is by far the longest. The disproportionate length of this part becomes understandable if one notes that it includes pestilential fevers as well as such systemic diseases as leprosy and boils, and also wounds. The six other parts deal, respectively, with the diseases of the head (II), the eyes, ears, nose and mouth (III), the throat (IV), the respiratory system (V), the liver, spleen, kidneys, and bladder (VI), and the reproductive organs (VII). Bernard thus covered the diseases in the traditional order from head to foot.[103] The seventh part, in which also such diseases as gout and varices are discussed, concludes with an antidotary in the head-to-toe order. Somewhat surprisingly, however, the final chapter of this antidotary returns to the head with a collection of cosmetic prescriptions for women.

Within almost all of the *Lilium*'s one hundred and sixty-three chapters Bernard rigidly adhered to a logical subdivision into six articles. First, he gave the definition, names, and kinds of the disease. He was mindful, however, of the Galenic principle that the knowledge of the real thing is more important than arguments about its names.[104] Where pertinent,

[100] "Hic traditur doctrina operandi, sed modus operandi particularis oportet frequenter mutare [Bernard here cites Heben Mesue]. Et ita diligens medicus poterit se aliquo modo in doloribus iuvare et informare": ibid., 1.21, fol. 12v. "Quoniam hic ponitur ars et doctrina, et multa relinquuntur medico particulariter operanti, quoniam in particularibus est magna diversitas": ibid., 1.19, fol. 11r. Also "componat igitur et permisceat cui datum est capud": ibid., 4.5, fol. 53r, and so on.

[101] "Hic non servavi ordinem solitum causa brevitatis": ibid., 6.15, fol. 88r.

[102] Diseases "curentur sicut dicetur in locis suis. Non enim esset decens quodlibet in quolibet loco ponere": ibid., 5.3, fol. 61v.

[103] This order seems to have originated with the *Historia naturalis* of Pliny (d. AD 79) according to Talbot, *Medicine in Medieval England*, p. 15. MacKinney, however, listed as the earliest instance the *Compositiones medicamentorum* of Scribonius Largus (first century AD), in "Medieval Medical Dictionaries," p. 243.

[104] "Sciencia medicine rerum est sciencia, non vocum, sicut scribitur 3. de morbo [Galen] et ideo nec de numero nec de nominibus curo, dum tamen res intelligamus. ... De nominibus igitur nulla est contencio, dum tamen res intelligamus": *Lilium*, 1.23, fol. 16r.

Bernard first described the anatomy of the affected organ, although his descriptions were much more detailed for external organs (eyes, ears, nose, and mouth) than for the viscera. In the next article he investigated the causes of the disease, "because if you do not know the cause, how can you effect a cure?"[105] He then followed with the diagnosis, or the description of the symptoms (*signa*), and the prognosis, or the forecast of the course of the disease. These two articles, the third and fourth in each chapter, were usually the shortest, although Bernard complained that "there are few physicians who know well how to diagnose,"[106] and he quoted Hippocrates' warning that "ugly errors" result from the physician's ignorance of the intensity of the disease.[107] The fifth article contained the prescribed treatment, which included one or all of the following: regimen and diet, therapy (ranging from phlebotomy to psychological shock), pharmacy, and – if all else should fail – surgery.[108] The sixth and last article of the *Lilium*'s chapters clarified difficult points and solved apparent contradictions between different authorities. The questions of this *clarificacio* are often academic and show indeed "traces of the scholastic method of instruction."[109] This is evident not only when Bernard uses the dialectical format of question–thesis–objections–answer–conclusion, but also when he occasionally ventures into such speculative problems as "whether sneezing is a natural or an unnatural movement."[110] Such instances, however, constitute only a fraction of the clarifying articles, and they hardly warrant the view that the entire *Lilium medicine* was "a faithful reflection of the subtle scholasticism of the age."[111]

[105] "In primis hic et in quolibet morbo intendimus ad causam, quoniam si causam ignorares quomodo curares": ibid., 1.7, fol. 4v.

[106] "Pauci sunt medici qui sciant bene distinguere": ibid., 4.5, fol. 52r. It may be noted that the term *distinguere* is apparently a latinization of the Greek διαγιγνωσκεῖν, whence the English "diagnose."

[107] "Et ideo dicebat Ypocras, neque quispiam medicorum mihi repertus est qui vires egritudinum cognoscere valeat ... et ideo fedi errores sequuntur": ibid., 4.12, fol. 59r.

[108] Bernard's treatment of surgery is discussed further in chapter 5 below.

[109] "Bernhard zeigt schon Spuren scholastischer Lehrmethodik in mancherlei akademischen Quaestiones...": Theodor Meyer-Steineg and Karl Sudhoff, *Illustrierte Geschichte der Medizin*, 5th ed. (Stuttgart, 1965), p. 146.

[110] "Queritur de sternutacione utrum sit modus morbi an sit motus nature": *Lilium*, 2.23, fol. 32v.

[111] "The work is a faithful reflection of the subtle scholasticism, the polypharmacy and the superstition of the age, and from its contents represents only a perpetuation of Arabic views...": Max Neuburger, *History of Medicine*, tr. Ernest Playfair (London, 1925), 2: 76.

The above allegation was taken over by several historians who also labeled the *Lilium* as a "characteristic Arabist textbook on the practice of medicine and typical of the scholastic subtlety and rigid adherence to dogma that was the mental attitude of the Middle Ages."[112] Since this verdict reflects on an entire era, it seems worthwhile to evaluate the *Lilium* in the perspective of its own period. A brief comparison may be made, therefore, with two closely related and near contemporary works, namely Gilbertus Anglicus' *Compendium medicine* or *Laurea anglica* (ca. 1250), and John of Gaddesden's *Practica* or *Rosa medicine* (ca. 1320).[113] The *Lilium medicine* is undeniably more "scholastic" than its two English counterparts, if by scholasticism is understood the logical organization of the subject matter which has been described above. The sevenfold division of the *Laurea* may have inspired Bernard, who improved on it by his systematic arrangement within each chapter. Thus, for example, Bernard placed his surgical references under the pertinent therapeutics, whereas Gilbert devoted some separate chapters to surgery.[114] John of Gaddesden, in his turn, apparently adopted Bernard's internal organization of the chapters, but failed to adhere to it consistently.

Besides its logical organization, other and more substantive features are attributed to scholastic medicine, such as the use of syllogistic reasoning and philosophical terminology, and references to Aristotle and the Salernitan Questions.[115] These features figure far more prominently in

[112] Donald Campbell, *Arabian Medicine and Its Influence on the Middle Ages* (London, 1926), 2: 163. This verdict seems based on Neuburger's convenient one-sentence evaluation, and was also borrowed, almost verbatim, by more recent historians. Their borrowings are poignant illustrations of non-medieval lack of originality: "It is a typical Arabist textbook ... characteristic of the Middle Ages, marked by scholastic subtlety and rigid adherence to dogma": Benjamin Gordon, *Medieval and Renaissance Medicine* (New York, 1959), p. 284. Also, in one of the most widely used manuals of medical history, we read that "The *Lilium* ... is a characteristic Arabist textbook..., nowise classic, and typical of the Middle Ages in scholastic subtlety and rigid adherence to dogma": Fielding H. Garrison, *Introduction to the History of Medicine*, 4th ed. (Philadelphia, 1960, [c. 1929]), p. 164.

[113] Gilbert, Bernard and John appear as a trio in Chaucer's *Canterbury Tales*: see chapter 1 above. I have examined the *Compendium medicine Gilberti Anglici* (Lyon, 1510) and the *Rosa anglica* (Pavia, 1492, Klebs 424.1), both in the New York Academy of Medicine. For a broader comparative study, see my article, "Scholasticism in Compendia of Practical Medicine, 1250-1450," *Manuscripta*, 20 (1976), 81-95.

[114] It is, therefore, misleading to assert that "Gordon omits the surgical passages which Gilbert incorporated in his work": Thorndike, *Magic*, 2: 480 and n. 1, quoting Henry Handerson, *Gilbertus Anglicus; Medicine of the Thirteenth Century* (Cleveland, 1918), p. 77, to the effect that "Gordon's work does not contain a single chapter on surgery proper."

[115] While the term "scholastic medicine" is widely used, an accurate definition on the

Gilbert's *Laurea* than in Bernard's *Lilium*. Gilbert also gave more weight to traditional authorities, since he intended primarily "to aggregate the particular sayings of major authors since they are most expert in the work of healing," and he preferred to let the reader choose from the different opinions presented.[116] Bernard, on the contrary, "quoted his authorities whenever he considered it expedient," and as Dr. Sarton aptly put it, "to bolster his own prestige."[117] Neither Gilbert's nor Bernard's procedure would seem to qualify for the charge of "rigid adherence to dogma." With regard to the charge of "Arabism" we may note that the authorities, cited in the *Lilium*, are predominantly classical: explicit references to Hippocrates (80) and Galen (226) outnumber those to Avicenna (130) and all the other Arabs (22) combined. Gaddesden's *Rosa*, on the other hand, which uses nearly five times as many citations as the *Lilium*, can be considered far more "Arabist," since Avicenna (with 474 citations), Haly (98), Averroës (98) and the other Arabs hold an impressive lead over Hippocrates (120) and Galen (417).[118]

It may be of interest further to note the view of recent historians of science and medicine, as well as those of medieval physicians, on the subject of the *Lilium medicine* and its counterparts. Dr. George Sarton claimed that the *Laurea*, *Lilium* and *Rosa* "represent three stages in the development of the general medical textbook of those times."[119] Our summary comparison, however, seems to indicate that in several respects the *Lilium* was an apex rather than an intermediary step in that development. This impression was shared by Dr. Paul Diepgen, who considered the *Laurea* a "compilation of traditions," and praised the *Lilium* for its "didactic and lucid presentation." He judged the *Rosa* "an

---

basis of a comparative investigation of the works would be a valuable contribution to scholarship. Meanwhile, a thoughtful and positive evaluation can be found in Paul Diepgen, *Frau und Frauenheilkunde in der Kultur des Mittelalters* (Stuttgart, 1963), pp. 82-89 and 157.

[116] "A morbis universalibus appositi nostri intentio est inchoare et dictiones particulares magistrorum quod in opere medicationis expertissimas aggregare": Gilbertus Anglicus, *Compendium medicine*, fol. 2r. And "consuetudo nostra est ex dictis meliorum meliora aggregare et ubi dubitatio est opiniones diversas interserere ut quisquis sibi eligat quam velit retinere": ibid., fol. 55v.

[117] George Sarton, "Lilium medicinae," in *Medieval Studies in Honor of J. D. M. Ford* (Cambridge, 1948), p. 240.

[118] The tally of quotations in the *Rosa* is given by William Lennox, "John of Gaddesden on Epilepsy," *Annals of Medical History*, 1 (1939), 284. Gaddesden also quoted Gordon three times. More comparative tabulations will be found in chapter 4 below.

[119] "Lilium medicinae," p. 242.

inferior imitation of the *Lilium*."[120] More significantly, this modern assessment apparently agreed with that of medieval physicians, such as Guy de Chauliac, who scoffed at "the fables of Gilbert" and the "faded English rose" of Gaddesden, but quoted some thirty times from Bernard de Gordon's *Lilium*.[121]

Almost as monumental as the *Lilium medicine*, by whose success it was inspired,[122] was Bernard's last dated composition, the *Liber de conservacione vite humane*, which he began in the Montpellier *Studium* on 22 February 1308[123] and apparently completed on 9 November of the same year.[124] This work comprises four books, which in manuscripts and printed editions have often been treated as separate treatises, namely *De flebotomia*, *De urinis*, *De pulsibus*, and *Regimen sanitatis*.[125] The author himself apparently realized that it was difficult to present as one work such disparate treatises whose subject matter ranged from therapeutics and diagnostics to hygienics. Therefore he endeavored to justify the unity of the entire *Liber de conservacione* by showing how its books were logically interrelated. This justification comprises part of his general prologue. His point of departure in planning the work is the discussion of

[120] The whole *Laurea* bears "den Charakter einer nur selten von eigenen Entwurfen unterbrochenen Zusammenstellung des Überlieferten...": Diepgen, *Frau*, p. 89. In the *Lilium* "imponiert die besonders klare Ordnung und lehrmassig gut verständliche kurze Darstellung des Gegenstandes": ibid., p. 92. The *Rosa* is "eine minderwertige Nachahmung des *Lilium medicinae*": ibid., p. 81.

[121] "Finalement s'est eslevee une fade Rose Angloise ... j'avois creu de trouver en elle suavite d'odeur, i'ay trouve les fables de l'Espagnol [Petrus Hispanus, author of the *Thesaurus pauperum*], de Gilbert et Theodoric...": Guy de Chauliac, *La Grande Chirurgie*, ed. E. Nicaise (Paris, 1890), pp. 14-15. At the same time, Guy judged that Bernard "valde bene tractavit" such a difficult subject as epilepsy (ibid., p. 414).

[122] See note 84 above.

[123] "Inchoatus est enim liber iste in preclaro studio montis pessulani anno incarnacionis domini [beginning 25 March] m°ccc°vij° xxij die februarij, luna celebrante synodum cum sole versus vij gradum piscium," *Liber de conservacione vite humane*, end of prologue, in Vatican MS Pal. 1174, first half 14th c., fol. 10r-v. Unless otherwise indicated, references hereafter are to this MS. The date is erroneously given as 1307 by Bell in "The Bibliography," p. 325. The dating and astronomical references are carefully verified by Ynez O'Neill in "The History of the Publication of Bernard of Gordon's Liber de Conservatione Vitae Humanae," *Sudhoffs Archiv*, 49 (1965), 269-279.

[124] "Inchoatus autem fuit iste liber xxii die Februarij sicut principio est dictum. Terminatus est autem cum auxilio magni Dei nona die Novembris eiusdem anni circulo, luna existente in virgine, vicesimo quinto gradu. Explicit liber de conservatione vite humane die innocencium [28 December] Mcccxi": *De conservacione*, 4: *Regimen sanitatis*, colophon, in London MS BL Sl. 3097, fol. 61r, cited by O'Neill, "The History," p. 273.

[125] See appendix below. The unity of the work was suggested by Sudhoff, "Schriftstellerei," pp. 171-174, and demonstrated by Bell, "The Bibliography," p. 328. Bell's article was brought up to date by O'Neill, "The History."

phlebotomy, to which the first book will be devoted – for a reason specified further in its own prologue, and from which the three other books will derive their *raison d'être* "because among all the tools with which the physician operates not one is found to work as marvelously as phlebotomy when it is done in the right way and in the proper place." In order to determine the correct procedure and the appropriate site, Bernard continues, we should know the patient's vital signs. Because this knowledge

> cannot be arrived at in any better way than by means of [examining] the urine, therefore I will collect some material on urines in the second book. Furthermore, because the amount of blood to be drawn could not be measured unless the [patient's] strength be known, and this strength cannot be known without the science of the pulse, therefore I will compile material on pulses in the third book.

Finally, stressing the fundamental dependence of therapy not only upon diagnosis but also upon hygienics, Bernard concludes,

> because phlebotomy pertains generally to the regimen of the human body, and the regimen cannot be known without the regimen of health, I will compose a regimen of health in the fourth book. Run, therefore, youths, run, elders, to this grand spectacle because every age, sex, and occupation will find here a pleasant and suitable abode in accordance with the nature and shape of his body.[126]

This part of the prologue has been quoted and paraphrased here at some length because it brings out two noteworthy points about the *Liber de conservacione*. First, it indicates that the entire work grew out of Bernard's special interest in phlebotomy – an interest that, as we shall see, was not merely academic. In addition, the excerpt illustrates the character not only of the general prologue but also of the prefaces to the individual books, a character that is perhaps the most striking feature of the work when compared with Bernard's earlier compositions. Besides glittering with subtle logic, the prologue and prefaces are much more elaborate and more speculative than anything Bernard had written to introduce his subjects of previous treatises, including the encyclopedic *Lilium*. One might interpret this feature as an indication of growing scholasticism, without overlooking a simpler explanation, namely that the author had become more contemplative with age. Although Bernard was perhaps only in his fifties

[126] *De conservacione*, prologue, in Vatican MS Pal. 1174, fol. 10r. The Latin text is also given by Sudhoff, "Schriftstellerei," pp. 174-175, and by O'Neill, "The History," p. 270.

when he composed the *Liber de conservacione*, his advancing age might also account for the heavily moralizing tone of his speculation. Thus, in the general prologue he admonished his readers to spend their allotted time span "entirely in virtuous deeds." This theme, presented in a gloss on Galen's lofty thought that "man is the prince and king of all animals and therefore walks with raised head,"[127] pervades the other prefaces.

Expanding on the idea of man's superiority, Bernard introduces the first treatise, *De flebotomia*, with the observation that "in the human body, Nature has enclosed blood as a treasure."[128] Then, however, after a brief description of the "digestion" of the blood as it was understood before the discovery of the circulation, he reveals how his study of phlebotomy and hematology – and thereby, as we have seen earlier, the entire *Liber de conservacione vite* – was prompted by a very concrete consideration. He relates that he has discovered in a certain group of people ("*apud quasdam gentes ... illius territorii*") blood with an off-white, ashen color. This discovery has greatly puzzled him because everyone in that group, "even those who [otherwise] fall within the range of health," had this condition. Therefore, he concludes, "I thought that there would be no better way for me to gain knowledge of this affair than by composing a treatise on phlebotomy." Lest we infer, however, that Bernard knew only the bookish road to the solution of a medical mystery, it should be underscored that he had also sedulously inquired, "*fui sedulus inquisitor*," about this problem with the professional "*flebotomatores*."[129]

In his preface to the second treatise, *De urinis*, Bernard returned to speculation. He again elaborated on an inspired Galenic thought, namely that "the search for the truth" is the most virtuous undertaking and therefore the highest road to God and to glorious beatitude.[130] He

[127] [*Incipit*] "Homo enim est princeps et rex omnium animalium et ideo incedit capite elevato: ista proposicio scripta est a Galeno in particula de iuvamentis membrorum in principio ... cavendum ergo est cum summa diligencia ut mensura temporis in virtuosis operacionibus totaliter expendatur": *De conservacione*, prologue, fols. 9v-10r.

[128] "In corpore autem humano natura reclusit sanguinem pro thesauro": ibid., fol. 10r.

[129] After describing his discovery, for which see chapter 1, p. 8, n. 36 above, Bernard states, "quod reputavi prodigium grande, cum ut plurimum in omnibus et sub latitudine sanitatis existentibus hoc sit verum, et consideravi quod non possem melius venire in cognicionem huius cause nisi quod ordinarem unum tractatum de flebotomia" (ibid.). "Investigacione huius coloris fui sedulus inquisitor apud flebotomatores" (ibid., fol. 20v).

[130] [*Incipit*] "Dilexi veritatis scienciam investigare, credens nullum melius posse adiuvari in mundo et magis proximus fieri Deo. Ista proposicio scripta est a principe medicorum Galeno vij de ingenio cap. primo. Innuit nobis Galenus modum et viam per quam possumus nos in via disponere ad beatitudinem. ... Quod autem beatitudo

demonstrated this idea with quotations from Aristotle, Scripture, and Saint Augustine. The general tenor of the preface, which is constructed as an intricate syllogism, is that bodily health, necessary to reach complete beatitude, cannot be known without the science of uroscopy. "In the major premise three subjects are to be expounded: virtuous works, free choice, and saving grace." After the major has been proven, "the minor remains to be explained, that the search for truth is a most virtuous undertaking." The conclusion, "it is clear that he who strives for truth directs himself towards happiness," in turn becomes the major premise for a second syllogism, deducing that "a good disposition of the body can not be completely known without the knowledge (*sciencia*) of urines."[131] The involved logic and the appeal to free choice and virtuous works to justify the usefulness of "collecting some material on urine" make Bernard's introduction to the *De urinis* seem incongruous and disproportionately elaborate.

By far the longest preface, however, is that which introduces the fourth and last part of the *Liber de conservacione*, namely the *Regimen sanitatis*. Here, skilfully blending Aristotelian, biblical, Galenic, and Augustinian ideas, and piously moralizing, Bernard described the three ages of man, first, until about fourteen, childhood, in which man leads "a life of pleasure"; next, until about thirty five, adolescence or youth, which is devoted to civic or political affairs; and, finally, old age, in which man leads a "contemplative or speculative life."[132] Bernard praised especially the third age and the contemplative or "divine or angelic" life as the happiest and most blessed, while also admitting that "old age is the path to death."[133] As if looking forward to the end of his own days, he concluded

---

adquiratur per inquisitionem veritatis ... potest manuduci per hunc modum. ... Studium veritatis est summa disposicio ad [ad]quirendum lumen beatitudinis gloriosum": *De conservacione*, 2: *De urinis*, prologue, fol. 26v.

[131] "In *majori* igitur proposicione tria sunt declaranda [viz. operacio virtuosa, liberum arbitrium, gracia gratum faciens]. ... *Patet igitur maior proposicio* ... restat igitur declarare *minorem*, quod autem studere in veritate sit opus valde virtuosum. ... Ex dictis igitur *patet* quod qui studet in veritate multum ad beatitudinem se disponit ... bona corporis disposicio non potest perfecte cognosci sine sciencia urinarum, ideo utile est multum aliquid de urina aggregare...": ibid., fols. 26v-27r (italics mine).

[132] [*Incipit*] "Secundum intencionem Aristotelis primo ethicorum vita humana est triplex, vita sc. voluptuosa et vita civilis seu politica et vita contemplativa seu speculativa...": *De conservacione*, 4: *Regimen sanitatis*, preface, fol. 52v.

[133] "Tercia autem etas est etas senectutis et in hac vita vivunt homines supra se quam dicunt vitam contemplativam seu speculativam seu divinam seu angelicam ... et ita senes sunt beati et felices. ... Et dicit Galenus senectus est semita ad mortem; mors autem secundum Aristotelem iii. ethicorum est ultimum terribilium; tota autem etas senectutis non est aliud nisi mors": ibid., fols. 53v-54r.

the preface with the following summation: "Let us begin and carefully follow these three ages from the day of birth to the final hour of death so that, after departing, we shall with God's help come rejoicing before the tribunal of the Judge."[134] He avowedly intended "to prolong somewhat" his *Regimen sanitatis*, in a leisurely mood that seemed lacking in his earlier treatises.

When we juxtapose the prefaces with the treatises themselves of the *Liber de conservacione*, we notice that the treatment of the subject matter is different in tone and tenor: it is still characterized by the conciseness, solid matter-of-factness and careful organization which we have come to expect of Bernard. The text of *De flebotomia* opens with an entire chapter devoted to the definition of phlebotomy which is gleaned from Avicenna.[135] The rest of the treatise is based primarily on Galenic works, especially his commentary on the Hippocratic *Aphorisms* and also his *De ingenio sanitatis*, to which Bernard often refers with the exact folio, column, and even line.[136] He may, in fact, have been referring to a standard codex that was available to the members of the university, especially since both these works were part of the list of required books, issued by Pope Clement v in 1309.[137] However, Bernard does not follow the sequence of the Galenic texts, and he complains that the rules for venesection were given "by Galen in a great number of places because it was Galen's habit rarely to assemble all [related] materials in one place." Therefore, and because the principles of phlebotomy have also been discussed in other standard works such as the *Almansor* of Rasis, the *Practica libri regum* of Haly and the *Canon* of Avicenna, Bernard judges that "it is necessary to assemble these rules in accordance with the teachings of the masters."[138] Some of the rules pertain to the blood-letting

[134] "Ut ergo huius corpusculi aliqualiter prolongare possimus regimen sanitatis cum dei auxilio inchoemus istas tres etates diligentissime persequendo a die nativitatis usque ad ultimam horam mortis et ita post separacionem cum dei auxilio gaudentes ante tribunal iudicis veniamus": ibid.

[135] [*Incipit*] "Flebotomia est evacuacio utilis multitudinem humorum evacuans: hoc est descripcio Avicenne fen iiij primi libri": *De conservacione*, 1: *De flebotomia*, fol. 10v.

[136] Thus, Bernard refers to the "vii° folio, circa dimidio" of Galen's commentary on the *Aphorisms*, but urges "et lege totum commentum" (ibid., fol. 15r). Also, "Galenus x de ingenio, cap. ix, linea xxxiij" (ibid., fol. 12v); and so on. Since the references occur also in other early MSS (e.g. Cues 308) and are an organic part of the text, it is unlikely that they were later interpolations.

[137] *Cartulaire*, document 25. Also in the 1340 curriculum: ibid., document 68.

[138] "Ista autem materia pertractata est a rasi vij° almansoris et quarto, et est pertractata ab haly ix practice libri regum et cum hoc ab avicenna primo canone capitulo de flebotomia, et insuper a galeno in multis locis valde, quoniam ista fuit consuetudo galeni

procedure, while others define the favorable and unfavorable times for venesection, the amount of blood to be drawn, and the particular veins to be struck. Still other canons prescribe the patient's diet during the time of phlebotomy. The latter point, however, is not developed by Bernard, "for the sake of brevity."[139] He intended also to be brief with regard to another topic related to phlebotomy, namely the diagnostic use of the drawn blood, for the stated reason that he "fears prolixity and it might perhaps be somewhat beside our purpose,"[140] a somewhat surprising statement in view of his initial reason for composing this treatise. Nevertheless, despite this disclaimer, Bernard devoted four chapters to such a diagnosis, which he found most useful "even though physicians are mostly negligent in inspecting the blood drawn by phlebotomy."[141]

More prevalent than phlebotomy as a traditional diagnostic technique was uroscopy, although in practice it apparently also had some opponents since Bernard scored the "arrogance and ignorance of those who disdain and loathe the judgments of urine."[142] His treatise *De urinis* depended heavily on the standard authorities, especially Isaac Judeus (fl. 900),[143] whose work on urines was included in the 1309 Montpellier syllabus,[144] as well as the two other "classics" of uroscopy, Theophilus (fl. 600) and Gilles de Corbeil (d. 1224). Bernard also quoted extensively from the Hippocratic collection, from at least fifteen Galenic works, and from Avicenna's *Canon*. This gave a very scholarly flavor to Bernard's book on urines, that was evidently adapted to his teaching. He gave special consideration to "the young," not only when he simplified the quantification of qualities, but also when he listed some special recipes.[145]

---

quod raro voluit omnino totam materiam aggregare in uno loco. Oportet igitur istos canones aggregare secundum intencionem doctorum": *De flebotomia*, fol. 12v.

139 "... et multa consimilia que cadunt circa istam materiam sed dimitto propter brevitatem": ibid., fol. 23v.

140 "Et si quodlibet istorum declareretur timeo prolixitatem et forte aliquo modo esset extra propositum. Ecce tamen quid dicit avicenna ... et galenus...": ibid., fol. 19v.

141 "Utilissimum igitur signum videtur esse sumptum a sanguinis disposicione facta per flebotomiam, licet medici ut plurimum sunt negligentes in aspiciendo sanguinem extractum per flebotomiam": ibid., fol. 20r.

142 "Patet arrogancia et ignorancia quorumdam qui dedignantur et nauseant iudicia urinarum," *De conservacione*, 2: *De urinis*, fol. 28r.

143 [*Incipit*] "Urina est colamentum sanguinis et similiter ceterorum humorum ex nature actionibus natum: ista proposicio scripta est ab Isaach versus principium libri de urinis": ibid., fol. 27r. There is a recent Italian translation by Silvano Chelini and Aldo Scapini, *Il 'De urinis' di Bernardo di Gordon (XIII-XIV sec.)*, (Pisa, 1968).

144 *Cartulaire*, document 25. Works on urines were also composed at Montpellier by Gualterus Agulinus (Agilon), Jordanus de Turre, and Albert of Montpellier.

145 "Brevior tamen et facilior modus graduandi propter iuvenes est iste.... Et hec

Furthermore, the didactic character of the *De urinis* may be responsible for an odd feature of its composition. That is, the treatise is made up of two dovetailing halves, each of which seems to have its origin in a different teaching approach. Most of the twenty-seven chapters fall into two parts which are often only slightly connected. The first part deals systematically with the elements and criteria of uroscopy, comprising the different colors, amounts, and substances of urines. In the second part of each chapter, however, Bernard either prescribed the treatment of a specific disease, not necessarily urinary, or he discussed an academic question. While this arrangement may seem dichotomous, it was evidently so planned since the author outlined the organization of his treatise in the prologue, where he proposed that to the discussion of uroscopy "we will add treatments and preventive methods and we will supplement whatever we omitted in the *Lilium*, especially with regard to semi-tertian agues, about which we have said little in the *Lilium* for the sake of brevity." In addition, he promised, "in each chapter we will, to the best of our ability, insert some question (*dubitacionem*)."[146] The initial question, "whether there could be a science of urine," is dissected with the scalpel of subtle logic: the answer is negative, if urine is considered as a subject by itself (*absolute*), but affirmative if it is seen *relative*, in relation to other diagnostic methods. As the final answer, however, he comes to the sound conclusion that our knowledge (*sciencia*) of urines should not be "vague and confused, but certain and specific."[147]

Bernard is not entirely specific with regard to another diagnostic tool, the pulse. His treatise *De pulsibus* is very brief, a mere two folio pages in the manuscripts, compared with twenty six folio pages for the *De urinis*; it is also rather amorphous – there are no chapters – and it is largely speculative. The basic definition of *pulsus*, for example, is explicated in a rambling literary discourse that likens the pulse to a herald who never lies and to a sentinel who never rests in revealing the unseen.[148] The discourse

---

[graduacio] sufficit propter iuvenes, quoniam exquisite et in puncto tradita est doctrina in libello de graduacione": *De conservacione*, 2: *De urinis*, fol. 35v-36r. "Et quia iuvenes multum gaudent in experimentis, ideo aliqua ponamus, Rx...": ibid., fol. 42v.

[146] "Hic addemus curaciones et preservaciones et supplebimus si que obmisimus in lilio et specialiter de emitriteis de quibus parum diximus in lilio causa brevitatis ... in quolibet capitulo inseremus aliquam dubitacionem iuxta modulum notre potencie": ibid., fol. 27r.

[147] "Posset aliquis dubitare utrum de urina posset esse sciencia. ... Quare concluditur quod non sufficit habere scienciam de urinis vagam et confusam, sed certam et determinatam, sicut erit inferius per totum cum dei auxilio manifestum": ibid., fol. 28r.

[148] [*Incipit*] "Pulsus est nuncius qui non mentitur, preco qui occulta reserat motibus

also includes a synopsis of the philosophical proposition that the heart is the seat of life, for which Bernard cites Aristotle, Galen, Augustine, Avicenna, and Alfred of Sareshel.[149] The major theme of the treatise is that the science of pulses, however useful, is so difficult that "only God, Galen his servant, and a very few others" are privy to it.[150] Among the reasons for this difficulty are cited, without further elaboration, the prerequisites for an accurate reading of the pulse, namely a high intelligence, a keen sense of touch, a knowledge of musical proportions,[151] and above all a long experience. In spite of his own many years of practice, however, Bernard appeared unable to formulate a list of specific instructions. The only concrete items in the *De pulsibus* are a sketchy anatomy of the heart and an elementary classification of the types of heartbeats. Bernard concluded this tract apologetically, "I do not attempt to go into the details of pulses because I have collected from the book on pulses of venerable Egidius as much as my weak competence would allow." And he advised readers, who were not satisfied with his own work, to consult that book of Egidius or Gilles de Corbeil.[152] One may wonder whether Bernard's "weak competence" reflected not merely the inherent difficulty of reading the pulse but also a lacuna in his own training, since he did not cite the treatises on pulses by Philaretus and Theophilus, which were part of the standard collection of medical textbooks, known as the *ars medica* or *articella*, but were absent in the Montpellier curriculum.[153] It is also worth noting that Bernard had waited for thirteen years after first expressing his plan to compose a treatise like

---

contrariis disposicionem cordis et arteriarum et tocius corporis manifestans. In ista descripcione tangitur tota natura pulsualis...": *De conservacione*, 3: *De pulsibus*, fol. 51r.

[149] "Alvaredus scripsit quod cor est vite domicilium": ibid., fol. 51v. The reference is probably to the *De motu cordis*, which Alfred of Sareshel (early 13th c.) dedicated to Alexander Neckam (1157–1217). For Alfred see Thorndike, *Magic*, 2: 187 and passim; also Sarton, *Introduction*, 2: 561-562 and passim.

[150] "Constriccionem pulsus solus Deus novit et Galenus servus eius et quidam alii paucissimi": *De pulsibus*, fol. 52r.

[151] In ignoring this opportunity for a discussion of the relation between pulse and music, Bernard differs markedly from several other medical authors, such as those examined by Nancy C. Siraisi, "The Music of Pulse in the Writings of Italian Academic Physicians (Fourteenth and Fifteenth Centuries)," *Speculum*, 50 (1975), 689-710.

[152] "Ad particularia autem pulsuum non condescendo quoniam secundum ultimum mee debilis potencie aggregavi supra librum venerabilis Egidii de pulsibus et cui istud non sufficit ad illud transvolvatur": *De pulsibus*, fol. 52r.

[153] *Articella* (Venice, 1487), at the New York Academy of Medicine. See *Cartulaire*, documents 25 and 68 for the curricula of 1309 and 1340 respectively. Two treatises on pulses were composed at Montpellier around the middle of the thirteenth century: one by Magister Cardinalis, the other by Gualterus Agulinus.

this, though apparently more elaborate than it eventually turned out. In his *Liber pronosticorum* (1295) he shortened the chapter on the pulse with the excuse that "we intend, with divine help, to discuss it elsewhere in the form of tables.... I say only this about pulses, that he who has the art and long practice will be less ignorant about pulses. Excuse me now from a more elaborate description, because with God's grace I will attempt to tackle this arduous matter more fully."[154]

The *Regimen sanitatis*, too, had been announced by Bernard long before its actual composition. In the *Liber pronosticorum* (1295) he had promised to say more on the ages of man "in the book on the regimen of health"; and in the *Lilium medicine* (1303–1305) he had outlined the highlights of hygiene "even though we intend to publish a longer treatise on the regimen of health."[155] Now, in 1308, this treatise became the last but most important part of the entire work, for which it supplied the title, *De conservations vite humane* or *On the Preservation of Human Life*. In the *Regimen* Bernard drew on the wisdom of his long experience as a teacher and physician and he adorned his instructions with the fruits of his wide reading. He cited proportionately more authors than he had in his other works, although he still placed the emphasis on Galenic writings, especially the *De regimine sanitatis*. Incidentally, this work of Galen did not appear in the Montpellier curriculum until 1340.[156] It would seem that the study of the rules of health as a separate subject was still in a formative phase at Montpellier in the beginning of the fourteenth century. The first general regimen had apparently been written there in 1281 by an unknown author.[157] In 1298 Arnold of Villanova composed a *Regimen sanitatis* after he had perhaps edited the *Regimen sanitatis Salernitanum*.[158] The following year a regimen by Avenzoar (d. 1162) was

[154] "... quoniam alibi cum auxilio divino intendimus pertractare per modum tabularum. ... De pulsibus dico tantum quod qui habebit artem et exercicium longum minus de pulsibus ignorabit. Excuso me nunc a prolixiori determinacione, quia per dei graciam largius temptaturus sum hanc materiam arduam aggredi": *Liber pronosticorum*, 4.5, in Vatican MS Pal. 1174, fols. 90v-91r.

[155] "Sicut melius dicemus in libro de regimine sanitatis, cum regimen principaliter pertractabimus": ibid., 2.6, fol. 83r. And "licet per dei graciam de regimine sanitatis longiorem intendamus edere tractatum": *Lilium medicine*, 5.8, fol. 66r.

[156] *Cartulaire*, document 68.

[157] See TK 313, item 12. The author may have been the "bachelor of Montpellier" of another regimen, listed in TK 731, item 11. The general regimen, in its turn, became the basis for custom-made regimens for individuals, produced by Montpellier masters such as Arnold of Villanova, Gregory, Jordanus de Turre, and others.

[158] While traditionally it has been believed that Arnold at least wrote a commentary on the Salernitan *Regimen*, Karl Sudhoff suggested that Arnold was in fact the editor of

translated by Profatius Judeus with the cooperation of the surgeon Bernardus Honofredi at the request of Master Petrus de Capitestagno.[159] Bernard de Gordon does not mention any of these works, although his reaction to some of their theses might be revealed in a textual comparison. While he "dealt with the regimen of health better than Arnold of Villanova" in the opinion of a fifteenth-century reader (who did not know his identity),[160] Bernard himself did not criticize or even acknowledge his colleague's 1298 *Regimen*. It is equally surprising that he cited neither the *Regimen Salernitanum*, which he had quoted several times in the *Lilium medicine*,[161] nor Avenzoar's work, although he knew its translator Profatius.[162]

More significant than these omissions are the weaknesses apparent in the organization of the *Regimen*. While Bernard had set out to discuss the life regimen from birth to death,[163] he did not carry the chronological development beyond youth, and he failed to outline a regimen for middle or old age. He, nevertheless, showed concern for old age when he raised the question "whether one could be preserved by the art of medicine ... so that he would be long-lived and would reach the limits of old age."[164] This question occurs, oddly enough, in the chapter on "the expulsion of superfluities." This brings us to another organizational flaw in the *Regimen*. As he had done in the book *De urinis*, Bernard inserted after several of the chapters questions that are rather speculative and often little related to the topic at hand. For example, it seems incongruous that the chapter on diseases of infants is followed by a question about the

---

the treatise itself: see "Zum Regimen Sanitatis Salernitanum," *Archiv für Geschichte der Medizin*, 12 (1919), 158-160. The latest expert on Arnold, on the contrary, classified even the commentary among Arnold's spurious works: Juan Arellano Paniagua, *Estudios y Notas sobre Arnau de Vilanova* (Madrid, 1963), p. 42. For the date of Arnold's own *Regimen*, see ibid., p. 73.

[159] See Wickersheimer, *Dictionnaire*, 1: 76.

[160] "Iste valens dominus et magister qui compilavit presentem librum non nominat in hoc volumine nomen suum. Revera circa sanitatem corporis ipse melius ordinavit quam magister Arnaldus de Villanova": 15th-cent. colophon in Paris MS BN 16190, fol. 40v.

[161] See chapter 4, note 16 below.

[162] Bernard had requested Profatius to draw up an armillary, as we saw in chapter 1, p. 10 above.

[163] [*Incipit*] "Mulier cum approximatur ad partum debet ire suaviter per loca declivia...": *De conservacione*, 4: *Regimen sanitatis*, fol. 54v.

[164] "Queritur utrum aliquis possit per artem conservari in regimine sue sanitatis sine errore ita quod esset longevus et veniat ad terminum senectutis et videtur quod sic": ibid., fol. 72v.

"complexion" of wine.[165] A further shortcoming of Bernard's arrangement relates to the material covered by the chapters of the *Regimen*. While Bernard asserted that "the regimen of health lies in the due administration of the six non-natural things,"[166] that is air, food and drink, excretion and repletion, sleep and wakefulness, exercise, and "accidents of the mind," he did not systematically develop these six essentials. Thus, he omitted a discussion of air but inserted chapters on massage, sex, and fatigue. Since the paradigm of the "non-naturals" was apparently well established at the time,[167] it seems unlikely that Bernard's inconsistency was due to the fact that he had only recently become familiar with the doctrine. Yet, such an explanation is suggested by his omission of these six essentials in the *Lilium*'s chapter on the regimen, where instead he had listed seven criteria for the preservation of health.[168]

It would perhaps be presumptuous to draw direct and definite conclusions from the above omissions and organizational flaws. However, these do suggest that Bernard was less thorough in the *Regimen sanitatis* than in most of his earlier treatises. His advancing age does not seem to be an adequate excuse, since the first two parts of the *Liber de conservacione*, composed within the same half year, were still remarkably thorough. Rather, Bernard may again have been handicapped by what he earlier termed "weak competence," or a lacuna in his own educational background, as we observed with regard to the preceding treatise *De pulsibus*. In general it appears that not only the academic treatment of the regimen of health was still in a developing stage at Montpellier, as we have seen, but also that the subject appears to have been neglected by practising physicians. Bernard himself charged that "it is most shocking that physicians of our time do not care to know the regimen of health because they do not consider it lucrative, whereas they are quite concerned and careful in the treatment of diseases and especially of fevers."[169]

[165] "Queritur cuius complexionis sit vinum, et mittitur quod humidi": ibid., fol. 57r. Another example, "Queritur utrum balneum aque frigide impediat augmentum," after the chapter on "accidencia anime," ibid., fol. 69r.

[166] "Regimen sanitatis consistit in debita applicacione sex rerum non naturalium": ibid., fol. 72v.

[167] See my "Theory and Practice," p. 106 and n. 9. I hope shortly to examine further the fate of this paradigm in Montpellier writings around 1300.

[168] "Circa regimen sanitatis cadunt VII consideraciones, prima sumitur a quantitate cibi et potus, secunda a qualitate, tercia ab ordine, quarta a complexione, quinta a consuetudine, vi. a tempore, vij. ab etate": *Lilium medicine*, 5.8, fol. 66r.

[169] "Ymmo quod est magis abhominabile medici nostri temporis non curant scire

Even if a certain weakening in presentation may be discernible in the books of his *Liber de conservacione*, this does not eclipse Bernard's continued efforts to supplement the teaching and to correct the practice of his contemporaries. In the survey of his indubitably authentic and precisely dated writings, one feature has become progressively clearer, namely the gradual evolution in the style, content, and method of his exposition. His writing career, which spanned over a quarter of a century, appears to have reached a turning point in 1303, the year of his *De gradibus* and, especially, of his *Lilium medicine*. In the early part of his career Bernard adhered closely to the Montpellier curriculum in choosing the topics of his treatises. In his later years, without losing sight of the curricular framework, he ventured into relatively new territory, even though the lacunae in his own background apparently hampered him in the achievement of his desired goal. Furthermore, his early works were consciously brief, primarily practical, and rigidly organized, while several of his later works, on the contrary, were more drawn out and more speculative. They were also weaker in organization and were often marred by inconsistencies. Finally, while in the beginning of his career Bernard mentioned few of the authorities he used, with the *Lilium* and subsequent works he became more explicit in his references. It is impossible to determine with certainty whether the developments in Bernard's writing reflect merely his personal evolution or were due to the wider context of growing interest in scholasticism at Montpellier. Nevertheless, these developments may guide us in the evaluation of his undated works as well as of the treatises attributed to him, to which we shall turn in the following chapter.

---

regimen sanitatis quia videtur eis quod non sit ibi lucrum, sed in curacione morborum et potissime febris sunt bene solliciti et attenti": *Regimen sanitatis*, fol. 72v.

3

Undated and Possibly Spurious Writings

Less widely known than those works of Bernard de Gordon which have been examined in the preceding chapter, but equally integral and important parts of the Gordonian corpus are his two monographs, *De tyriaca*[1] and *De marasmode*.[2] These monographs constitute a closely related pair; they are found together in at least one very early manuscript,[3] and they may have been composed in quick succession. Both treatises, though undated in the manuscripts, appear logically to belong to the last period of Bernard's writing career, and they have several significant characteristics in common with the *Liber de conservacione vite humane*. Most striking are the unprecedented length and speculative nature of the prologues. The preface of the treatise *De tyriaca* constitutes the elaborate first chapter of the work, and that of the *De marasmode* occupies almost one third of the entire treatise. Each preface, moreover, deals with an academic *questio*, in which Bernard avowedly observed his customary procedure. He began the tract *De tyriaca* with the definition,

[1] Ed. Julius Pagel, "Über den Theriak, nach einer bisher ungedruckten Schrift des Bernhard von Gordon," *Pharmaceutische Post*, 27 (1894), 205-206, 220-224, 235-237, 257-262. I am grateful to Professor McVaugh, who provided me with a copy of this edition, which is very difficult to find. References hereafter are to this edition, which I have found to be at minimal variance with one of the oldest MSS of *De tyriaca*, London BL Sl. 3096, "13-14th c." (TK 1574), fols. 337r-338v.

[2] I used photocopies of two early MSS: Vatican Pal. 1234, early 14th c., fols. 135v-139v, to which further citations refer; and London BL Sl. 3096, early 14th c., fols. 335v-336v. The former is not in TK; the latter is, though not attributed: TK 848, item 6. Since this treatise has never been printed and is extant in only a few manuscripts, I owe special gratitude to Professor Kibre, who discovered the Vatican MS and drew my attention to it in 1970. I hope to publish shortly a text edition from several manuscripts, together with a detailed analysis of the *De marasmode*. See also n. 28 below.

[3] London MS BL 3096: see preceding notes 1 and 2. As may be observed there, the *De marasmode* precedes the *De tyriaca*, although I am discussing them here in reverse order for the sake of clarity.

> Theriac is the great preparation that cures all poisoning on account of its property. This definition we may now expound in six ways, in accordance with our custom: first, by doubting.... First tractatus, on the exposition of the question whether theriac can be administered to healthy bodies....[4]

The treatise *De marasmode* opens more abruptly,

> In order that we may observe our custom before we come to the subject, let us inquire about something that relates to this matter, and the question is, whether seminal substances materially enter the constitution of the fetus.[5]

The phrase *consuetudo nostra*, conspicuous in both prefaces, would seem to indicate that Bernard had already arrived at an habitual pattern of teaching and writing, unless the term referred to an established scholastic procedure in use at the Montpellier faculty of medicine. It is possible that the scholastic *dubitatio* was especially important in the discussion of such special subjects as theriac, the universal antidote, and marasmus, or consumption by desiccation. These subjects apparently did not find a specific place in the general curriculum. They nonetheless lent themselves well to philosophical debate in the context of courses on medical theory.

The two above named tracts, *De tyriaca* and *De marasmode*, do indeed contain more philosophical speculation and medical theory than Bernard's earlier works. There is also more emphasis on Aristotelian concepts such as causality, matter and form, and act and potency. In both treatises, moreover, Bernard cited Averroës as well as Aristotle with a frequency unequalled in his other writings, and he also gave more numerous explicit references to the customary authors, including Galen and Avicenna. Not only the range of his citations, but also the format and the choice of subjects indicate that with both monographs Bernard intended to supplement the contemporary university curriculum. These features seem to fit into the pattern of his later, rather than earlier, years of composition.[6]

[4] [*Incipit*] "Tyriaca est confectio magna curans omnem venenosam aegritudinem a proprietate. Istam enim diffinitionem possumus juxta quandam nostram consuetudinem exponere sex modis: 1° dubitabiliter. ... Tractatus primus de expositione quaestionali (Quaestio). Quaeritur ergo utrum tyriaca possit exhiberi corporibus sanis...": *De tyriaca*, 1, pp. 220-221.

[5] [*Incipit*] "Sed ut servemus consuetudinem nostram antequam veniamus ad intencionem queramus aliquod circa istam materiam et erit questio, Utrum seminaria intrent materialiter constitucionem fetus": *De marasmode*, prologue, in Vatican MS Pal. 1234, fol. 135v. References hereafter are to this manuscript, unless otherwise indicated.

[6] Compare also the synoptic table of Bernard's authorities in chapter 4 below.

Hence, we are led to the more significant conclusion that the two works seem to clinch the pattern of Bernard's evolution. They may also have reflected contemporary trends at Montpellier.

For these as well as other reasons the two monographs should be dated as precisely as possible. It is safe to assume that they were composed after 5 February 1305, since Bernard did not include them in the *Lilium*'s epilogue, where he methodically listed his previous *opuscula* which, "if added to this work [the *Lilium*], would constitute the *opus* of my complete *magisterium*."[7] The assumption that 1305 is the *terminus a quo* for the composition of the *Tractatus de tyriaca* finds further support in two passages of the *Lilium medicine*. First, in the chapter on snakebites, Bernard mentioned theriac among the antidotes but did not describe its composition; and in the same chapter, in a paragraph on the viper (*tyrus*), "from which theriac is made,"[8] he said nothing about his detailed description of the preparation of viper troches, found in the *De tyriaca*. In the latter, he asserted that the viper troches, together with troches of squill and *diacorallus*, were to be blended with honey and forty-nine simples to make theriac. The second indication that the *De tyriaca* was composed after 1305 is found in the *Lilium*'s chapter on poisons, where Bernard adopted the theory that theriac should not be administered to healthy bodies. This view, which contradicted that of Avicenna, was expressed by Averroës in his own tract *De theriaca*. Bernard in the *Lilium* also repeated the crux of Averroës' argument, borrowed from Galen, that antidotes are substances between poisons and medicines.[9] Here again Bernard made no mention of his own *De tyriaca*, although he devoted the introductory *questio* of this latter work entirely to Averroës' proposition. Omissions like these would be very surprising if the *Tractatus de tyriaca* had been composed before the *Lilium medicine*. And this would be especially so for an author who consistently made cross references to his own pertinent discussions.

[7] "Intelligendum est hic quod nos ordinavimus per dei graciam quatuor opuscula diu est propter istud opus, et si adderentur hoc esset opus perfecti magisterii, scilicet libellus de regimine acutorum et libellus de crisi et de diebus creticis, et libellus qui intitulatur tabula ingeniorum, et sine ista factum est nichil, et libellus de graduacione," *Lilium medicine*, epilogue, in London MS BL Harl. 3698, fol. 99v.

[8] "... tyri, ex quibus conficitur tyriaca": ibid., 1.14, fol. 8r.

[9] "Intelligendum quod medicine liberantes a venenis secundum Galenum et Averroym verbo suo de tyriaca sunt medicine inter medicinas et venena ... tyriaca tamen non competit in corporibus temperatis ... ista sunt verba Averroys in libro suo de tyriaca": ibid., 1.13, fol. 7v. This passage is missing in the three printed editions of the *Lilium* which I have examined.

In fact, the treatise on theriac is apparently his only indubitably authentic work that neither mentions nor is mentioned in any other of his writings. This exceptional feature led Karl Sudhoff to believe that the *De tyriaca* was an "early work of his youth, perhaps the earliest extant composition" of Bernard de Gordon. Sudhoff suggested further that the more mature master had come to think of it as a "mere compilation," and therefore did not deem it worthy to be counted among his full-fledged works.[10] However, the distinction between compilation and original composition is based on modern criteria rather than on those of Bernard and his contemporaries.[11] Also, if the *Tractatus de tyriaca* had been considered an inferior work by the author's contemporaries, it would probably not have been preserved in several copies. At least nine manuscripts of the tract are extant, and it is twice cited in a work as famous as Henri de Mondeville's *Chirurgia*.[12] Henri de Mondeville began his *Chirurgia* in 1306, a date which provides us with a *terminus ad quem* for the completion of Gordon's *De tyriaca*. It may also be noted that Henri de Mondeville did not cite the *Epistola de dosibus tyriacalibus* of his better known contemporary, Arnold of Villanova. Perhaps Henri de Mondeville, as a surgeon, found Arnold's thoroughly philosophical *Epistola* less useful than Bernard's simpler treatment of the subject.[13]

Although it seems somewhat surprising, too, that Bernard did not refer explicitly to Arnold's *Epistola*, which was presumably written before 1300,[14] it is possible that he intended to supplement the work of his colleague. More specifically, in his *Tractatus de tyriaca* Bernard may have attempted to bring Arnold's esoteric discussion down to a more accessible

[10] "Ich halte diese ... kleine Abhandlung aber für eine frühere Jugendarbeit des Gelehrten ... vielleicht seine früheste erhaltene Ausarbeitung." In a footnote Sudhoff added that Bernard seems "sie dann allerdings als pure Kompilation gar nicht zu seinen 'Werken' gerechnet zu haben." Ironically, on this assumption Sudhoff then based an evaluation of Gordon as "progressive": "Er hatte sich damit also von mittelalterlich literarische Hochwertung rein kompilatorischer Arbeit in bemerkenswerter Weise emanzipiert": Sudhoff, "Die Schriftstellerei," p. 180 and n. 2.

[11] See, for example, the observations made in the preceding chapter with regard to the *Tractatus de gradibus* and the *Lilium*.

[12] *Die Chirurgie des Heinrich von Mondeville nach Berliner, Erfurter und Pariser Codices*, ed. Julius Pagel (Berlin, 1892), pp. 406 and 474.

[13] For a stimulating analysis of the difference between Bernard's approach and that of Arnold with regard to theriac, and of its implications for Montpellier's academic medicine "from a still somewhat empirical to a markedly philosophical or theoretical orientation," see Michael McVaugh, "Theriac at Montpellier, 1285-1325 (with an edition of the *Questiones de Tyriaca* of William of Brescia)," *Sudhoffs Archiv*, 56 (1971), 113-144.

[14] Arnold's *Epistola* "was finished early in Arnald's tenure as master at Montpellier between 1292 and 1299," according to McVaugh, "Theriac," p. 115.

level, as he had done earlier for the question on medicinal degrees.[15] This is suggested in Professor McVaugh's observation that, while Arnold's *Epistola* aims at "rational understanding," Bernard's *Tractatus* emphasizes "factual information" and displays a "disinterest in abstract matters."[16] As we shall see, however, Bernard's "disinterest" might also be characterized as a certain inability to deal with abstractions, rather than as indifference. In his tract on theriac he seems to have been tantalized by philosophical speculation, but he apparently became entangled in a maze of inconsistencies and sophistries. Thus, while he stressed, on the basis of Galen and Avicenna, the paramount role of experience in determining the effect of theriac,[17] he utilized purely deductive reasoning in his introductory *questio* whether theriac could be administered to healthy persons. Furthermore, while adopting the negative answer by which Averroës contradicted Avicenna, Bernard ventured the apologetic but unsubstantiated suggestion that Avicenna could not have been so mistaken.[18] At the same time he ignored Averroës' cogent philosophical arguments, and he substituted a rambling syllogism on the theory that both man and the universe are governed by the interplay between lack and abundance. To that theory Bernard then applied the Aristotelian three kinds of justice (*legalis*, *distributiva*, and *commutativa*) to prove that theriac can only harm the healthy body. Throughout, while groping for an understanding of natural philosophy, he seemed unable here to reach beyond the levels of, at best, ethics and dialectic or, at worst, literature. We may, therefore, agree with the comment of the modern editor of the treatise, Dr. Julius Pagel, that the first chapter meant "a definite retrogression" in Bernard's composition and that "it all sounds very profound but is actually most uninspired."[19]

Subjects such as the basic reason for theriac's effectiveness, which could have drawn on the insights of natural philosophy, receive only brief and vague treatment in the central chapters of the *Tractatus de tyriaca*.

[15] The possible relation between Bernard's *Tractatus de gradibus* and Arnold's *Aphorismi de gradibus* was discussed in chapter 2, pp. 49-50 above. However, whereas Bernard supported Arnold's views in the *De gradibus*, he implicitly disagreed with Arnold on theriac. Bernard followed Averroës, whom Arnold – an almost fanatical anti-Averroist – denounced.

[16] McVaugh, "Theriac," pp. 116 and 118.

[17] "De aliqua virtute [tyriace] penitus nihil potest sciri nisi solo experimento sicut dicit Galenus in de simplici medicina cap. 2 in fine capituli et Avicenna in V° canonis": *De tyriaca*, ed. Pagel, p. 257.

[18] "Credo quod non fuit intentio Avicennae quod tyriaca competeret in corpore bene sano, sed in notabiliter lapso": ibid., p. 223.

[19] Pagel, "Über den Theriak," pp. 205 and 224.

Instead, Bernard elaborated upon the history of theriac's invention and he developed a gloss on its definition. In that exegesis he also inserted an interesting excursus on various poisoning techniques, with a certain Polys as his source of information,[20] and he sternly warned against the use of poison, citing Seneca, Sallust, Suetonius, and the Hippocratic Oath.[21] He further added a chapter on the usefulness of theriac in warding off more than a dozen diseases, ranging from migraine to pestilence, but he admitted that "although theriac was valuable in antiquity, in modern times it has little effect on the body because of its fast action."[22] Bernard's skepticism in this regard may explain why he was rather casual in determining the dosage of theriac, which was to be "conjectured in approximation to the truth."[23] He nevertheless recounted an instance in which he had corrected a near-fatal overdose which an apothecary had taken:

> It happened that a certain apothecary had sold a toxic substance and some remained behind the nail of his thumb. When he began to pick his nose with his thumb, his color began to change and he fainted repeatedly. When he took three drams [of antidote], his condition became more serious. I was called around midnight and I saw that the amount of theriac had been too high. I made him vomit and after he had purged his stomach sufficiently, I gave him one dram of theriac and he was cured.[24]

The entire matter of theriac's ambivalent effect, however, became ironically academic when, in the last chapter of the *De tyriaca*, Bernard gave the recipe "of his own invention" for a panacea that, unlike theriac, was not itself poisonous.[25] The universal potency of that panacea, moreover, exceeded that of theriac and, in fact, surpassed "all natural

[20] Perhaps Pollex ('Poli medici'), *De venenis, incipit* "Prima pars est de speciebus venenorum..." (TK 1094).

[21] "Dimisit nobis Hippocrates in mandatis in testamento suo: non doceas potionem mortiferam nec aborsum": *De tyriaca*, ed. Pagel, p. 259.

[22] "Valebat igitur thyriaca antiquitus sed moderno tempore propter velocitatem actionis in corpore parum valet": ibid.

[23] "Dosis autem tyriacae quaedam conjectatur proxima veritati ad minus potest esse 1/2 scrup. sed in fortioribus venenis dr. 2. Alia autem quae sunt inter ista sunt doses mediae": ibid., p. 260.

[24] "Accidit quod quidam apothecarius vendiderat toxicum et remansit in ungue pollicis aliquid et incepit fricare nares ab intus cum pollice et incepit color ejus mutari et frequenter syncopizavit et accepit dr. 3 et magis gravabatur et versus mediam noctem fui vocatus et vidi, quod quantitas tyriacae erat nimia et feci quod evomeret et postquam sufficienter evomuit, dedi sibi dr. 1 tyriacae et curatus est": ibid.

[25] "Nichilominus possibile esset per naturam altiorem invenire confectionem et magis amicabilem corpori humano quam sit tyriaca ubi nulla est venenositas sicut in fine huius libelli manifestabitur": ibid., p. 223.

understanding." This prescription, Bernard claimed further, was "marvelously beneficial to old people."[26] He even went a step further, both in his avowed consideration for the aged and in his promise of marvelous results, when he elsewhere prescribed an electuary that, "with a good regimen, prevents man from growing old, in so far as that is possible in nature."[27] This electuary he placed in the last chapter of his *Tractatus de marasmode*, a chapter that shows intriguing similarities with the final chapter of the *Tractatus de tyriaca*.

Bernard's concern with old age is only one among several characteristics of his later writing period. Other characteristics were his growing penchant for speculation, to which he devoted lengthy prefaces; his interest in philosophical questions, which he evidently failed to integrate fully into medical theory; and his endeavors to explore topics that were presumably omitted in the university curriculum at Montpellier. These features of Bernard's composition were only faintly foreshadowed in the *Lilium medicine*, but they became progressively more pronounced in the *Liber de conservacione vite humane* and in the treatise *De tyriaca*, and they apparently converged in his work *De marasmode*. For this reason the *Tractatus de marasmode*, though probably the only authentic and original work of Bernard's that has never been printed or analyzed on its own terms,[28] deserves to be dated, at least tentatively, and to be described, at least briefly, in this study.

[26] "Cap. sextum de medicamine alciori quod excedit tiriacam et omnem perspicacitatem luminis naturalis [interlinear addition: id est secundum aliquas species diptami [?] ipse tamen exponit in lilio capitulo 13 de venenis]. Rx... Istud autem electuarium in causis calidis et frigidis et in egritudine et in omni causa ... valet et in passione cuiusquamque membri ... et mirifice prodest senibus.... Et est de invencione nostra. benedictus igitur qui terminavit et menti aride tribuit roritatem. Amen": *De tyriaca*, in Vatican MS Pal. 1083, fol. 326v. The Latin text of this chapter was omitted in the edition of *De tyriaca* by Pagel, who published it in his *Angebliche Chirurgie des Heben Mesue Junior* (Berlin, 1893), p. 144.

[27] "Capitulum ultimum de confeccione ducente ad insenescibilitatem. Colligantur in primis quinque aque. ... Istud enim electuarium supposito bono regimine reddit hominem insenescibilem quantum est possibile per naturam quoniam confortat cor et viscera ... et addit in calorem naturalem ... quoniam ex *alciori* potencia (sciencia) processit quasi sit qualibet *perspicacitas luminis naturalis*. Et *est de invencione nostra. Benedictus igitur qui* incepit terminat et menti (*menti aride*) *tribuit* veritatem (*roritatem*). *Amen*": *Tractatus de marasmode*, fol. 139v. The variants (in parentheses) are found in London MS Bl. Sl. 3096, fol. 336v. Emphasis mine, in order to highlight the parallels with the passage in the preceding note. The similarities were also pointed out by Sudhoff, who published this chapter (from Erfurt MS Amplonianum Q.227) without knowing that it belonged to the *De marasmode*: "Schriftstellerei," pp. 186-187.

[28] See note 2 above. The *Tractatus de marasmode* was mentioned in passing by Sudhoff, who thought that it might be a chapter of the *Lilium medicine*: "Schriftstellerei,"

As was the case for the *Tractatus de tyriaca*, our earlier assumption, based on the epilogue of the *Lilium medicine*, that the *Tractatus de marasmode* was composed after 1305, is also borne out by some further indications. In the *Lilium*'s chapter on sterility Bernard inserted a brief *clarificacio* of the role of semen in conception.[29] Given his method of cross-references to his own more specialized discussions of a topic, it is significant that at this point he did not refer to his *De marasmode*, whose introduction revolves entirely around the same subject, as we shall see shortly. A similar argument may be used to move the date of composition after the completion of the *Liber de conservacione vite* in November 1308. In his prologue to its fourth part, the *Regimen sanitatis*, Bernard briefly described life as a slow death through the "unceasing consumption of our radical moisture by our bodily heat,"[30] without referring to the text *De marasmode* of which this was the central theme. The *incipit* of this text supplies a perhaps even stronger clue for the latest possible date of composition. As Professor McVaugh has recently pointed out,[31] the opening lines seem to correspond, not with the early Latin translation of Galen's treatise *De marasmo*, but to the second translation by Niccolò da Reggio, who was active from about 1308 to 1345.[32] Since so many indications point to a late date of composition for Bernard's *Tractatus de marasmode*, we may probably assume that the treatise reflects the thoughts and concerns of his more mature years.

The most salient feature of the *Tractatus de marasmode* is the author's preoccupation with questions of natural philosophy that are related to medicine. In the long introduction he examines the question "whether semen materially enters the composition of the fetus."[33] He asserts that the philosophers, namely Aristotle and Averroës, answer the question negatively. But they are contradicted by the physicians, namely Hippocrates, Galen, Avicenna and Haly. As could be expected, Bernard takes the side of the latter group. He also adds the somewhat sarcastic

---

p. 184. See also Pagel, "Über den Theriak," p. 205. Some important aspects of the treatise were most recently discussed by Michael McVaugh, "The 'Humidum Radicale' in Thirteenth-Century Medicine," *Traditio*, 30 (1974), 274-278 and 281.

[29] *Lilium medicine*, 7.14, fol. 95r.

[30] "Creati enim sumus in ultimo humiditatis ... et non cessat calor nostri corporis consumere humidum radicale donec mors adveniat": *De conservacione*, 4: *Regimen sanitatis*, fol. 53v.

[31] McVaugh, "The 'Humidum Radicale'," p. 274.

[32] See Lynn Thorndike, "Translations of Works of Galen from the Greek by Niccolò da Reggio (c. 1308–1354)," *Byzantina Metabyzantina*, 1 (1946), 226.

[33] "Utrum seminaria intrent materialiter constitucionem fetus": *De marasmode*, fol. 135v.

comment, "if one, therefore, wishes to be productive (*fecundus*), let him take the way of the physicians (*viam medicorum*); but let him, who wishes to speculate instead, take the path of the philosophers (*viam philosophorum*)." And he further cites Avicenna's remark that "in this case, ignorance of philosophy does not harm the physician. For, whether this or that be said, generation occurs anyway."[34] Nevertheless, Bernard endeavors not only to sum up the philosophers' propositions but also to answer them briefly in a philosophical fashion.

The major arguments for the Aristotelian thesis, that semen plays no material but only a formal role in conception, are summarized by Bernard as follows: first, the sperm relates to the fetus as the sculptor to the statue, the leaven to the bread, and the rennet to the cheese; second, something cannot be at the same time a formal and a material principle; and third, the agent (or active cause) and matter (or passive material) cannot coincide. He replies *ad primum* that not the sperm's form, whose role is admittedly analogous to those of the sculptor or the rennet, but its matter enters the composition of the fetus. *Ad secundum*, he simply reiterates the distinction between the seminal form and the sperm's form-receiving matter which becomes part of the conceptus. *Ad tercium* Bernard argues that the agent and matter do not coincide because "the seminal matter that was previously under the form of the sperm is now under the form of the fetus."[35] Thus, in a few deft strokes he disposes of the Aristotelian position whose defense occupied several chapters in works of natural philosophy such as the *De animalibus* of Albertus Magnus.[36] Yet Bernard claims that his proof has the strength of Achilles and that his reasoning is as good "a demonstration as can be made regarding this matter."[37] It is further worth

[34] "Qui ergo vult esse fecundus viam teneat medicorum; qui autem aliter voluerit speculari teneat viam phylosophorum. Avicenna autem dicit in canone [3. *fen* 20] quod ignorancia phylosophie in hoc casu non nocet medico. Si autem dicatur sic vel sic, nichilominus fit generacio": ibid., fol. 136r. For the original formulation of the "philosophers' thesis" see Aristotle, *Generation of Animals*, 1.18-21 and 2.4. For the "physicians' thesis" see Galen, *Usefulness of the Parts of the Body*, 14.2.

[35] "Ad tercium similiter, unde agens et materia non coincidunt quia perit forma ... et ita materia seminalis que tunc erat sub forma seminis nunc est sub forma fetus": *De marasmode*, fol. 136r.

[36] Albert's views on the roles of male and female sperm are summarized in my forthcoming paper (with Anthony A. Travill, M.D.) on "Human Embryology and Development in the Works of Albertus Magnus." The preparation of this paper has yielded insights which I hope to apply to a more thorough and comparative analysis of Bernard's opening *questio* in the essay that will introduce my edition of the *De marasmode*.

[37] "Et ista videtur demonstracio sicut potest fieri demonstracio circa talem materiam, et ideo in ista materia ista racio videtur Achilles, cuius operaciones erant eroyce et immortales secundum Aristotelem": *De marasmode*, fol. 136r.

noting that, while his arguments follow the hylomorphic terminology of the philosophers, they really hinge upon a premise of medical theory, namely that a zygote can develop into a fetus only from the "proper mixture" (*temperies*)[38] of the thick male sperm and the watery female "sperm" and that, therefore, both the male and the female contributions are materially necessary for conception.[39] His argumentation indeed also refutes the Aristotelian view that women do not contribute true sperm, but it does not clarify the logical connection between this secondary proposition and the main question on the role of semen in conception.

Another logical connection not clearly established by Bernard is that between his introductory *questio* on semen and the subject proper of his *Tractatus de marasmode*. He does assure us that the question is *circa istam materiam*,[40] pertinent to the topic covered by the title, and the connection is presumably implicit in the theory that semen, which is "the residuum of the fourth nutrient,"[41] is also the source of the "radical moisture," which is a key element in the discussion of marasmus.[42] Without unfolding this connection, however, he abruptly moves from the introductory *dubitatio* to the subject matter of the treatise, marasmus, which is defined as "the decay of the living body by desiccation."[43] This opening definition was taken, as we have indicated, from Galen's work *De marasmo*, on which Bernard's treatise was a commentary, apparently the only one composed at Montpellier and perhaps even the only one dating from the Middle Ages.[44] The core of Bernard's commentary, however, namely the concept of the "radical moisture" or *humidum radicale*, was an idea that saw "its most active efflorescence just before and just after the

[38] Compare pp. 87-88 below.

[39] "Semen viri est magis grandinosum et spissum et non est bene extensibile per se. Semen autem mulieris est magis aquosum et ideo quando coniunguntur resultat quadam temperies; quare utrumque est necessarium": *De marasmode*, fol. 135v.

[40] See note 3 above.

[41] "Semen autem quod est residuum quarti cibi...": *De marasmode*, fol. 136r.

[42] For the connection between semen and radical moisture in Avicenna's thought, see Thomas S. Hall, "Life, Death and the Radical Moisture: A Study of Thematic Pattern in Medieval Medical Theory," *Clio Medica*, 6 (1971), 4.

[43] "Capitulum primum. De diffinicione marasmodis et eius explanacione. [*Incipit*] Marasmon est corrupcio viventis corporis in siccitate": *De marasmode*, fol. 136v. So far as I can determine, the spelling *De marasmode*, from the Greek *μαρασμόν*, seems to have been used only by Bernard, while others preferred the latinized form *marasmus*.

[44] Bernard's treatise is "secundum intencionem Galeni" according to the title in the manuscripts. Galen's *De marasmo* was also known, sometimes as *De tabe*, more often as *De marcore*: ed. C. G. Kühn, in *Claudii Galieni Opera Omnia*, 7 (Leipzig, 1824), pp. 666-704.

beginning of the fourteenth century."[45] It would seem that, with his timely study of a difficult topic, Bernard intended to supplement the Montpellier curriculum, as he may have done earlier in his *De pulsibus* and in his *Regimen sanitatis*. It is also possible that in his *De marasmode*, as in his tracts on degrees and on theriac, he was again inspired by the work of his philosophically inclined colleague Arnold of Villanova. In his treatise *De humido radicali*, composed before 1308, Arnold had concentrated on the question of life in terms of natural philosophy rather than in direct relation to medicine.[46] Professor McVaugh has recently pointed out that, because of the contrast between the endeavors of Arnold and those of Bernard, "it is difficult to assign proper credit to the intellectual consistency and philosophical sophistication of Bernard's thought."[47] To this observation may be added the comment that Arnold often seemed to pursue an abstract subject for its own sake, whereas Bernard examined such subjects primarily in the light of medical practice. However, if Bernard had entirely lacked interest in abstraction, it would be difficult to understand why he devoted so much of his treatise *De marasmode* to the theoretical aspect of the radical moisture and its drying out.

The doctrine of the radical moisture revolved around the belief that this moisture was the fourth of the "secondary humours" or "humidities" in the human body. This theory had gradually been grafted onto the Hippocratic-Galenic tradition of the four basic humours in the body, namely blood, phlegm, yellow bile or choler, and black bile or melancholy.[48] Its rationale was based on the need to explain the phenomena of life and death; and its philosophical background was epitomized in Bernard's description of the two complementary "principles of life," the natural or innate heat and the radical moisture.[49] With regard

[45] Hall, "Life, Death and the Radical Moisture," p. 10.

[46] I consulted Arnold's *De humido radicali* in the printed edition of his *Opera* (Lyon, 1509), fols. 38v-42v. For manuscripts see TK 413 and 415. Arnold's treatment of the radical moisture is ably analyzed by McVaugh, "The 'Humidum Radicale'," pp. 278-281; his ideas were also surveyed briefly by Hall, "Life, Death and the Radical Moisture," pp. 10-14.

[47] McVaugh, "The 'Humidum Radicale'," p. 277.

[48] For a survey of the Greek roots and Arabic (especially Avicennan) formulation of the doctrine, see further the essential article by Hall, who seems to have opened up the topic for exploration. For Montpellier before 1300, see further McVaugh's study.

[49] "Capitulum quartum de manifestacione marasmodis per etatem senectutis ... duo sunt principia vite, scilicet calidum et humidum. ... Quamdiu autem hoc calidum naturale potest esse in humido radicali, tamdiu est vita. ... Et ideo nichil aliud est mors quam extinccio calidi naturalis in humido radicali": *De marasmode*, fol. 137v.

to the interaction between heat and moisture, Bernard also devoted an entire chapter to the traditional comparison between the decaying human body and the burnt out wick of a lamp.[50] Although the two vital principles as well as the wick simile had appeared sporadically in Bernard's earlier works,[51] it was only in his *Tractatus de marasmode* that he attempted to correlate them in a systematic discussion. However, the task seems to have been beyond Bernard, either because the whole doctrine was still in a chaotic developing stage or because he was once more unable to carry through a process of abstraction. Apparently confused by the differences between the basic Galenic formula and that of Avicenna, he explicitly stated that "there are three humidities in our body,"[52] but actually listed four, namely *ros*, *cambium*, *glutinum*, and *humidum radicale*. Moreover, he gave two discrepant sets of these secondary moistures: in one set, for example, he omitted the *humidum radicale* but added "hard and soft fat (*pinguedo et adeps*) and flesh" as well as a vague class of "humours in the veins and arteries."[53] In a further inconsistency that betrays his bewilderment, Bernard related each of the humidities to the specific kind of hectic fever that affected them, but then concluded that there is really no criterion to distinguish one kind of hectic fever from another.[54]

For the diagnosis of hectic fever as distinguished from continuous and

[50] "Capitulum tercium de manifestacione marasmodis per exemplum. Circa naturam marasmodis est advertendum ad ea que apparent nobis in lampade...": ibid., fol. 137r-v. For the use of this lamp simile by Galen and Avicenna, see Hall, "Life," pp. 5 and 7.

[51] Thus, in the *Liber pronosticorum*, 1.6, "De febrium differencia," fol. 80r; in the *Lilium medicine*, 1.9, "De febre ethica," fol. 5v; in the *De tyriaca*, ed. Pagel, pp. 222-223; and in the *De conservacione*, 4: *Regimen sanitatis*, Prologue, fol. 53v. However, in the *Lilium* Bernard mentioned only the "substancialis humiditas" and the wick, in the *De tyriaca* only the "calor naturalis," and in the *Regimen sanitatis* only the "humidum radicale." It is also worth noting that neither the term *marasmon* nor its synonyms are found in any of the above works, including the encyclopedic *Lilium*.

[52] "Capitulum quintum de manifestacione marasmodis per species ethicorum. Intelligendum igitur ... quod in corpore nostro est triplex humiditas": *De marasmode*, fol. 138v. With "certain pre-Avicennian Arabs ... Galen's trio became a quartet to which Avicenna presently gave the name 'secondary'," according to Hall, "Life," p. 7.

[53] The first set: 1. humores in venis et arteriis, 2. pinguedo et adeps et caro, 3. ros et cambium, and 4. glutinum: *De marasmode*, fol. 137v. The other set: "1. in concavitate venarum et arteriarum [London MS Bl. Sl. 3096, fol. 336v, has *capillarium*] ... et appellatur ros, 2. ... in porositatibus et foraminibus membrorum ... et appellatur cambium, et ex hac materia nutriuntur membra et residuum fit materia seminalis. 3ª autem humiditas appellatur glutinum, et hoc est cum cambium transubstanciatum est ad naturam membri...," and this set becomes more confused until Bernard gets to 4, the "humidum radicale": *De marasmode*, fol. 138v-139r.

[54] "Videtur michi quantum est de presenti quod in aliqua specie febris ethice non est aliquid signum distinctum distinguens realiter unam speciam ab alia": ibid., fol. 139r.

putrid fevers, Bernard directed the reader to his earlier discussion: "about this matter we have spoken with regard to Galen's book *On the Differences of Fevers*, and therefore I do not pursue it here."[55] Since a little further he called that discussion "a certain question on the book *On the Differences of Fevers*,"[56] it would appear that he had in mind a topical commentary on Galen's work. The same impression is strengthened by the terms used by Bernard in parallel references in his *Lilium medicine*[57] and, especially, in his book *De urinis*. In the latter, he only briefly summarized his solution of the contradiction between Galen and Avicenna on the cause of sanguinous fever because "this question has been settled by us at length with regard to Galen's book *On the Differences of Fevers*."[58] The precise Latin terms used, "*ista questio est determinata a nobis cum solemnitate supra librum Galeni de differenciis febrium*," strongly suggest a formal commentary. However, such a commentary has thus far not turned up in manuscript form. Bernard, moreover, did not refer to the *questio* with such terms as *commentum*, *liber*, *libellus*, *tractatus* or the like; and he did not include it among the *opuscula* listed, as constituting his complete *magisterium*, in the *Lilium*'s epilogue.[59] Hence, his discussion of the differences between fevers, which he cited so frequently, may not be a separate work but rather part of another Gordonian treatise, probably the sixth chapter of the first *particula* of the *Liber pronosticorum*.[60] Even though this chapter belongs to a work that in some ways might be considered a commentary on the Hippocratic *Prognostications* and it does not cite Galen by name, it appears to be the *questio* Bernard referred to in his later writings. The chapter, titled *De febrium differencia et pronosticacione in eisdem*, appears based primarily on the authority of Galen's *De febrium differentiis*;[61] it also examines the cause of sanguinous fever as well as the particular diagnosis of hectic fever, which were the subjects of Bernard's later cross-references as we have seen above; and, in addition, it contains a brief discussion of the "radical moisture" and the "natural warmth," the two principal themes of his *Tractatus de marasmode*.

[55] "De ista autem materia sumus locuti supra librum de differenciis febrium Galeni et ideo non curo hic [loqui]": ibid., fol. 138v.

[56] "De hoc [calore] dicetur in quadam questione supra librum de differenciis febrium": ibid., fol. 139r. The future tense (also in London MS BI. Sl. 3096) is somewhat puzzling in view of the past tense in the other *loci*.

[57] *Lilium medicine*, 1.5, fol. 3v; 1.9, fol. 6r.

[58] *De conservacione*, 2: *De urinis*, 20, fol. 43r.

[59] See note 7 above.

[60] In Vatican MS Pal. 1174, fols. 79v-80v.

[61] See C. G. Kühn, ed., *Claudii Galieni Opera Omnia* (Leipzig, 1821-1833), vol. 7.

Bernard's allusion, in the treatise on marasmus, to his own earlier discussion of fevers has been investigated here in some detail not only because it may prevent the assumption that he composed a separate work on fevers,[62] but also because it further illustrates the organic unity and systematic cohesiveness of most of his scholarly writing over a stretch of many years. The same observation applies to some additional instances where Bernard seemingly referred to other works by himself. Thus, in 1299, he announced that he was planning to write a separate treatise, "*singularem tractatum*," on leprosy;[63] but this treatise appears to have become incorporated as an elaborate chapter (1.22) in the *Lilium medicine*. In the *Lilium*, furthermore, he abridged his analysis of the question whether pain in itself is a disease, because the reader could consult his longer discussion of this issue in his commentary on the *Viaticum* (Constantinus Africanus' *Breviarium*).[64] In still another instance he apparently referred to his own commentary on the fourth book of Galen's *De morbo*.[65] In the case of the last two cross-references Bernard asserted that he had "discussed the matter," and he spoke of "our expositions." His use of the terms *disputavimus* and *exposiciones nostras* suggests the interesting possibility that he may have been alluding to commentaries which he had delivered in the course of class lectures. This would further underscore the academic orientation as one of the features of Bernard's medical writings.

The features of Bernard de Gordon's *opus*, which we have surveyed thus far, especially its close-knit unity and its gradual evolution towards speculation, make it difficult to accept his authorship for several works that have been attributed to him in manuscripts. Nevertheless, some of these works seem worth investigating because they were possibly either marginal or unfinished efforts by the older Bernard that would throw further light on his thought. Even if they were not from his hand, these

[62] For example, Silvano Chelini and Aldo Scapini, in the introduction to their translation, *Il 'De urinis' di Bernardo di Gordon* (Pisa, 1968), p. 7, list among Bernard's works a treatise *De febris* [*sic*] as composed in 1294.

[63] "Et de ista materia [viz. lepra] per Dei graciam singularem tractatum sumus facturi": *De decem ingeniis*, fol. 285r.

[64] "Avicenna et Geraldus supra Viaticum dicentes quod dolor est morbus ... moventur aliqui ad credendum quod dolor sit morbus per se; licet supra Viaticum et alibi multa dixerim de ista materia, tamen breviter aliqua dicam. ... Et quia disputavimus istam materiam supra Viaticum, ideo derelinquo": *Lilium medicine*, 2.10, fol. 26r. For Constantine's *Viaticum*, see TK, index, 1779; see also chapter 4, pp. 108-109 and nn. 26-29 below.

[65] "Qui autem causas passionis [sc. strabositatis] scire voluerit, habeat exposiciones nostras super iv de morbo": *Lilium*, 3.6, fol. 40r; also ibid., passim.

works may perhaps give us some idea of his image during the succeeding decades in which such writings were ascribed to him rather than to another author.

Of the possibly spurious treatises the one that has the strongest paleographical evidence in favor of Bernard's authorship is variously called *Libellus de impregnacione*, *Tractatus de sterilitate mulierum*, or *Tractatulus de sterilitate et eius cura*. It is attributed to Bernard in at least eight manuscripts that date from the early fourteenth century to 1458, and several of them are of Montpellier provenance.[66] In one of these, Paris MS Bibliothèque Nationale latin 7066, it appears in the company of related treatises by other Montpellier masters, such as the *De mulierum impregnacione* of Jordanus de Turre (fl. 1313–1335), the *De impregnacione mulierum* of Pierre de Nadilz (fl. 1369–1374) – who seems to have borrowed occasionally from our supposedly Gordonian tract – and *Contra difficultatem conceptionis* of Johannes Patarani (fl. 1375–1382). In another Paris manuscript, Bibliothèque Nationale latin 6957, the work is followed by the *De sterilitate* ascribed to Arnold of Villanova (ca. 1240–1311)[67] and then the *Secretarium practice* of Johannes Jacobi (fl. 1360–1380). In a third codex, Vatican Palatine Latin 1083, in which its transcription was completed on 22 August 1458 at Heidelberg, the *Tractatulus de sterilitate et eius cura*, "edited by Bernard de Gordon, dean of the distinguished Montpellier *studium*," occurs in the midst of authentic and attributed Gordoniana.[68]

While the paleographical data seem to argue for the attribution to Bernard of the *De sterilitate*, the textual evidence is far more ambiguous. This treatise is only one of more than a half dozen fourteenth-century writings that evince a lively interest in sexual reproduction at Montpellier but present a complex problem in the manuscripts. Some of these writings are anonymous while in other cases one treatise is ascribed to different authors; copies of the same treatise abound with variants as well as lacunae; and between different treatises there are numerous similarities in

[66] See appendix below.

[67] This *De sterilitate* is spurious according to Juan Paniagua, *Estudios y Notas sobre Arnau de Vilanova* (Madrid, 1963), p. 42. It was considered authentic, however, by Paul Diepgen for several reasons, especially because Gentile da Foligno (d. 1348) attributed it to Arnold: "Studien zu Arnald von Villanova," *Archiv für Geschichte der Medizin*, 6 (1913), 380-386.

[68] "Et sic est finis tractatus de sterilitate editus a bernardo de Gordonio decano preclari studii montispessolani. Finitus et scriptus per Johannem borner de bůdingen in octava assumpcionis beate marie Virginis anni etc. [14]58," Vatican MS Pal. 1083, fol. 321v. For other works in the same codex see appendix below.

form and contents. These similarities indicate either an intricate interdependence or the pervasive influence of a common source, itself apparently a digest not only of standard works such as Aristotle's *De animalibus*, Avicenna's *Canon* (Book 3, *fens* 20 and 21), and Trotula's *De passionibus mulierum*, but also of apocrypha such as the pseudo-Aristotelian *Problemata* (Book 4) and the *De secretis mulierum* attributed to Albertus Magnus.[69] To sort out the entire puzzle, of which the supposedly Bernardian treatise is a piece, would require a detailed comparative study of the texts and authors and thus would lead us too far afield. Suffice it here to examine briefly the *Tractatus de sterilitate* in relation to Bernard's authentic writings.

If Bernard did not himself write the tract on sterility, he probably earned its attribution to him by the elaborate chapters on the subject in his *Lilium medicine*, especially 7.1 and 14. The first noteworthy point about the *De sterilitate* is that it contains not a single explicit or indirect reference either to the *Lilium* or to any other work. This omission is especially striking in the instance of a fertility test or *experimentum*. In order to determine "whether a woman is ready to conceive," she should be suffumigated with aromatics, and the test is positive "if she smells the smoke through her mouth and nose" – to which one manuscript adds the somewhat sheepish warning, "be careful, however, that the woman have a good sense of smell, otherwise the test would be worthless."[70] In recommending this test, the author of the *De sterilitate* neither refers to the *Lilium*, where it is also found,[71] nor cites the source where it was introduced, whereas Bernard in the *Lilium* acknowledged that it had been "taught" by Hippocrates.[72] Some further discrepancies between the two discussions concern the presumed causes of sterility which indicate its treatment. For example, sterility "on account of shortness of the penis" is simply dismissed as irremediable in the *De sterilitate*,[73] unlike in the

[69] An excellent survey of the entire period is by Paul Diepgen, *Frau und Frauenheilkunde in der Kultur des Mittelalters* (Stuttgart, 1963).

[70] "Experimentum optimum ad probandum utrum mulier sit apta ad concipiendum aut non; suffumigetur cum gallia mustata et si senciat fumum per os et nares apta est et si non senciat non est apta": *Tractatulus bernhardi de gordonio de sterilitate et eius cura*, in Vatican MS Pal. 1083, fol. 321r. To this is added, "cave tamen quod mulier habeat bonum odoramentum, alias nichil valeret," in *Tractatus de impregnacione editus a magistro bernardo de gordonio*, in Paris MS BN lat. 6957, fol. [57r].

[71] "Ypocras enim docuit experimentum utrum mulier esset parata conceptui...": *Lilium medicine*, fol. 94v.

[72] See preceding note. The test is given in the Hippocratic *Aphorisms*, 5.59.

[73] "Si fuerit sterilitas propter brevitatem virge et quia foramen est tortuosum non recipit medicamen": in Vatican MS Pal. 1083, fol. 321r.

Lilium, where only "notable smallness" is admitted as incurable,[74] but for the problem of shortness are prescribed a coital knee-chest position for the woman and enlarging massages and plasters for the man, "as Galen teaches in *De ingenio*, towards the end."[75] Other causes, often blamed for failures in the mysterious reproductive process, were magical in nature. On this subject, too, the *Lilium* and the *De sterilitate* diverge: while Bernard in the former mentions only the *maleficium* or evil spell responsible for impotence,[76] the author of the latter suggests that for sterility "on account of spells (*vaticinia*, *fascinaciones*) one piece of silver be put under the head" of the sleeping or unaware partners.[77]

Although our treatise and the *Lilium* chapters diverge on some causes, they do agree in their adherence to Hippocratic-Galenic principles. Both discussions not only consider a general humoral imbalance as the prevalent cause of sterility,[78] but they also direct their treatments mainly towards the causes rather than towards the symptoms of the problem. Thus, the *De sterilitate* opens with a truism, "Since sterility occurs on account of a cause, let us therefore begin with that cause," and goes on to declare that "the temperateness of complexion is the cause of fertility, or [there may be] a slight difference, as long as the seeds meet each other in contrary qualities. Hence, disproportionate warmth is a cause of sterility, and it is known from a fast pulse...."[79] This declaration agrees with two basic ideas expressed also in the *Lilium medicine*. In the first place, there

[74] "Notabilis parvitas virge non recipit curacionem": *Lilium medicine*, 7.14, in Paris MS BN lat. 6964, fol. 76v.

[75] "Contra sterilitatem propter brevitatem virge, eleventur crura mulieris intantum quod semen possit cadere in fundo agri nature, vel magnificetur virga et verberetur suaviter cum virgis et emplastretur cum pice ut docet Galenus de ingenio versus finem": *Lilium medicine*, 7.1, in London MS BL Harl. 3698, fol. 89r. Compare Bernard's *De marasmode*, where Avicenna is cited as stating that "brevitas virge reddit hominem sterilem, quoniam non potest fundere semen ad fundum agri vite," in Vatican MS Pal. 1234, fol. 136v.

[76] "De paucitate coitus sive de frigidis et maleficiatis sive de sterilitate viri": *Lilium medicine*, 7.1, in London MS BL Harl. 3698, fol. 88v.

[77] "Si fuerit sterilitas propter vaticina [*sic*] ponatur ipsis ignorantibus argentum unum sub capite pacientis," in Vatican MS Pal. 1083, fol. 321r. "... propter fascinaciones ponatur ipsis dormientibus argentum unum sub capite in testa avellini [?]," in Paris MS BN lat. 7066, fol. 31v. The sentence is defective in Paris MS BN lat. 6957, fol. lvijr.

[78] See the Hippocratic *Aphorisms*, 5.62, ed. Kelly, p. 312.

[79] [*Incipit*] " Quia causa sterilitatis accidit ideo ab ipsa inchoemus. Temperamentum complexionis est causa fecunditatis vel modicus lapsus dum tamen obvient sibi semina in contrariis qualitatibus. Caliditas igitur nimia est causa sterilitatis, quod [*sic*] cognoscitur ex velocitate pulsus," in Vatican MS Pal. 1083, fol. 320r. For variant *incipits* see appendix below.

was the generally held belief that both partners must be "*temperati*" for fruitful intercourse.[80] The second idea, which is implicit in the above declaration, is that women also have sperm: this opinion was more controversial and had been defended by Bernard against an authority as venerable as that of Aristotle.[81]

The fact that no controversy was alluded to in the treatise *De sterilitate* is understandable since the author's intent was not to explain but to prescribe cures. Except for the references to "cause," "temperate complexion," and symptoms in the opening statement cited above, the treatise contains virtually nothing on etiology, humoral pathology, or diagnostics. It is impressively exhaustive, on the other hand, in the realm of therapeutics, and it runs through the entire spectrum of methods for the treatment of sterility. Among those methods suffumigations and pessaries figure most prominently, but the external remedies vary from phlebotomy and steam and sitz baths to ointments, plasters and poultices, and the oral prescriptions comprise a cornucopia of powders, pills, syrups and electuaries. In addition, dietary rules dictate specifically which wines, vegetables and viands are to be used or avoided, and even the psychosomatic approach is not neglected. Thus, "if sterility results from obesity of the uterus, one should procure emotions (*accidencia anime*) such as anger and worry as well as numerous vigils and a hard decubitus."[82]

The wide range of therapeutic methods in this treatise is almost matched by the diversity of *materia medica*. Although there are only two mineral ingredients, naval pitch and salt, the vegetable components are more numerous, including not only the ubiquitous rosewater and almond oil but also frankincense and myrrh, and the tips and seeds of poppies as well as opium. The richest variety, however, is in the category of animal ingredients, most of which are evidently chosen for their supposedly sympathetic powers. Thus, the fertile hare yielded to the concoctions its uterus, its heart, the rennet of its stomach, its droppings, and even its hairs, while the sparrow's potency seemed to rest in its brains. The most obvious examples of animal organs used for their sympathetic effective-

[80] *Lilium medicine*, fol. 95r. The cause seems to become effect in the *De marasmode*: "Quando [semina] coniunguntur resultat quadam temperies," in Vatican MS Pal. 1234, fol. 135v.

[81] *Lilium medicine*, 5.8, *Clarificacio*, fol. 92v, and *De marasmode*, fols. 135v-136v.

[82] "Si fuerit sterilitas propter pinguedinem matricis procurentur accidencia anime ut ira, sollicitudo, vigilie plurime et duri accubitus," in Vatican MS Pal. 1083, fol. 321r. This passage closely parallels the treatment prescribed in the *Lilium medicine*, 7.14, fol. 94v.

ness are the testicles of the fox and the penis of the bull. As was often the case in medieval pharmacopoeia, several ingredients were also credited in the *De sterilitate* with additional strength because they were rare and exotic, such as the fat of a lion or the uterus and the stomach rennet of a camel – although the author here realistically added, "if it can be found."[83] Not only exotic but also precious were the simples in a preparation admittedly reserved "for the rich: Rx two ounces of ambergris, one ounce of balsam, one half ounce of aloe wood, one dram of musk, and three ounces of calamint storax...."[84] The most expensive ingredients, such as file-dust of pierced pearls, pure gold and silver, hyacinth sapphires, and coral and emeralds, are prescribed in an individualized *consilium* that occupies almost one fourth of the treatise in several manuscripts.[85]

If we could determine the identity of the noble patient who was supposed to benefit from this *consilium* or *ordinacio*, namely "the Countess *Putemensi*, who was utterly barren (*quasi stupida sterilis*) from a cold or humid cause,"[86] we might perhaps gain a valuable clue to the origin of the entire treatise *De sterilitate*. At this time Bernard's authorship remains doubtful since possible arguments, based on the number of manuscripts and on some parallels with the *Lilium medicine*, are not only inconclusive but also weakened by several inconsistencies. Should this therapeutic treatise be authentically Bernardian, it would deviate from the pattern of his previously examined writings, in which we have consistently found a combination of practical advice and theoretical discourse. It is worth noting that nearly all other works attributed to Bernard also lack that combination and belong distinctly either to the category of theory or to that of practice.[87]

Of those other works only one is theoretical, and it does not go beyond a vocabulary of diseases with their etymology and classification. In its apparently single extant copy, which precedes the treatise *De sterilitate* and follows Bernard's *Tractatus super pronostica* [*sic*] in Vatican MS

[83] "Rx. coagulum cameli si inveniatur," in Vatican MS Pal. 1083, fol. 320r.

[84] "Item pro divitibus, Rx ambre grisee uncias ij, balsami unciam i, ligni aloes unciam et mediam, musci drammam i, storacis calamitis uncias iij...": ibid., fol. 320v.

[85] "[Rx] margaritarum perforatis [*sic*] limature auri puri et argenti an. drammam i, saphirorum iacinctorum corallis smaragdi an. drammas ij": ibid., fol. 321v.

[86] "Ordinacio facta pro comitissa putemensi que erat quasi stupida sterilis de causa frigida vel humida": ibid., fol. 321v.

[87] The only two exceptions, in which theory is combined with practice, are the *Codicillus seu De morbis et eorum curationibus* (see n. 93 below) and the *Breve compendium de neutralitatibus decidentie* (see n. 96 below).

Palatine 1083, the vocabulary bears the title *Arbor egritudinum*. After an introductory poem on the idea that a tree is known by its fruits, the text opens with a definition of disease as "a bad quality against nature, which perceptibly and directly harms the natural operations."[88] The diseases are then classified according to the conventional scheme of most compendia, from a basic division into general (systemic) and particular (local) to the ultimate subdivisions, among which the distinction between the three kinds of hectic fever is the most elaborate.[89] The peculiar branching procedure as well as the title of the *Arbor egritudinum* may have been inspired by similar arrangements, such as Ramon Lull's *Arbor scientiarum* (1296).[90] In any event, this synopsis of diseases illustrates the scholastic concern with logical organization that was also demonstrated in Bernard's writings. A related concern with etymology as a key to the understanding of reality is at the heart of this glossary. The author's principal sources are the lexicographers, especially Isidore of Seville (d. 636, *Etymologiae*), Papias the Lombard (fl. ca. 1050, *Lexicon*), Uguccio of Pisa (d. 1212, *Derivationes*), and Januensis (Joannes Balbus of Genoa, d. 1298, *Catholicon*; or Simon de Cordo of Genoa, fl. 1285–1295, *Synonyma medicine* or *Clavis sanationis*), all of whom he quotes by name.[91] Since the

[88] [*Incipit*] "Grata laboranti labor omnis bravia confert. ... Egritudo morbus seu infirmitas est mala qualitas preter naturam sensibiliter et immediate ledens operaciones naturales," in Vatican MS Pal. 1083, fol. 308v.

[89] "Et hec febris [ethica] habet tres species secundum triplicium humorum consumpcionem unde si consumitur primus humor membrorum qui vocatur ros tunc est prima species. Et hic humor roralis habet se in corde et membris sicut oleum in lampade; si consumitur cambium, qui humor habet se sicut oleum imbitum licinio vel licmini [*sic*] tunc est secunda species. Si consumitur tercius humor scilicet gluten qui se habet in membris solidis sicut consubstancialis humiditas ipsi lichinio tunc est tercia species: versus, Rorem prima species sequens dat cambia tercia gluten. Quidam distinguunt has tres species ethice sic. Quia vel est consumpcio humiditatis substancialis antequam tercia digestio est incepta et tunc est prima species qua facilis est curabilis. Vel est consumpcio humiditatis illius quando tercia digestio est incepta sed nondum completa, tunc est secunda species qua difficulter curatur. Vel est consumpcio predicte humiditatis quando tercia digestio est incepta pariter et completa, tunc est tercia species qua simpliciter est incurabilis": ibid., fol. 309v. This distinction is quoted here in its entirety because it relates to Bernard's discussion of hectic fevers and radical moisture (see notes 50-57 above) and it seems to summarize the earlier Montpellier interest examined by Professor McVaugh (see n. 28 above).

[90] For Lull's *Arbor scientie* as well as his *Arbor philosophie amoris* (1298) and *Arbor philosophie desiderate* (no date), see Sarton, *Introduction*, 2: 907.

[91] Other authors quoted include Galen, Alexander (of Tralles, fl. ca. 500) and Mundinus. By the latter was probably meant Mondino Friulano (d. 1340), author of an addition to Januensis' *Clavis sanationis*, rather than the more famous Mondino de' Luzzi (d. 1326). Medical encyclopedias were surveyed, with emphasis on alphabetical rather

author also cites Johannes de Tornamira (d. ca. 1396, *Introductorium*) and even Bernard de Gordon ("*Gordonius in Lilio*"),[92] the attribution of the treatise to Bernard is obviously an error of the scribe or rubricator. That error, of a kind not uncommon in manuscripts, was probably caused by hasty inference from the immediately preceding *explicit* of Bernard's work on prognostications, but it may have been induced by his systematic arrangement and etymology of diseases in the *Lilium medicine*.[93]

As in the case of the *Arbor egritudinum*, the attribution to Bernard of several other writings rests on very limited and doubtful manuscript evidence. Even so, a glance at some of these writings, all of which pertain exclusively to practical medicine, may enhance our understanding of Bernard's scholarship. It is quite instructive, for example, to compare his treatise *De urinis*, examined in the preceding chapter, with the brief *Opusculum de authenticis urinarum* [*signis*] *compositum per magistrum Bernardum de Gordonio*.[94] The latter deals only with diagnosis and prognosis based on a five-fold distinction in urines.[95] The treatise *De urinis*, on the contrary, went beyond these confines of uroscopy and not only covered aspects of physiology, such as the origin and function of urine, but it also raised questions of methodology, for instance "whether there can be a science of urine,"[96] and it offered several dietary and medicinal prescriptions "because the young are so fond of remedies

than on topical arrangements, by Loren C. MacKinney, "Medieval Medical Dictionaries and Glossaries," in *Medieval and Historiographical Essays in Honor of James Westfall Thompson* (Chicago, 1938), pp. 240-268.

92 "Gordonius in Lilio ponit tres species [melancolie] quarum prima generatur de causa frigida," in Vatican MS Pal. 1083, fol. 317v.

93 A much larger glossary of diseases, called *Codicillus seu De morbis*, which also contains therapeutic advice, is ascribed to Bernard de Gordon in Oxford St. John's College MS 197, 14-15th c., fols. 127r-254r. Its arrangement follows the conventional head-to-toe order, unlike the branching of the *Arbor egritudinum*, and it ends with fevers, unlike both the *Arbor* and the *Lilium medicine*, which begin with fevers.

94 In Oxford MS Bodl. canon. misc. 455, 15th c., fols. 239r-242v. Also in Cambridge Trinity College MS 500, 15th c., fols. 1v-3, according to TK 1609: I have not yet examined the latter manuscript.

95 [*Incipit*] "Urina quinquies distinguitur, scilicet ratione substantie, ratione turbulentie vel claritatis, ratione quantitatis, ratione colorum, et ratione contentorum," in Oxford MS Bodl. canon. misc. 455, fol. 239r.

96 See chapter 2, n. 147 above. Aspects of physiology and questions of methodology are combined with uroscopy and therapeutics in the *Breve compendium de neutralitatibus decidentie*, which was "editum per magistrum B. Gardonio [*sic*]" according to its only extant copy in Wolfenbüttel MS Aug. F.12.4, AD 1452, fols. 171-175. This compendium, together with a similar one ascribed to Johannes de Tornamira (ibid., fols. 168-171), was edited and analyzed in the "Inaugural-Dissertation" of Erna Margarete Kollert (Leipzig, 1924).

(*experimentis*)."[97] This work was composed for a younger, less experienced public and in close relation to teaching, as we have noted earlier, whereas the *Opusculum* was apparently intended for convenient consultation by the practitioner. Another striking difference between the two works lies in their use of sources, especially since both contain almost exactly the same number of citations. Unlike Bernard's *De urinis*, the *Opusculum* does not mention Hippocrates; it cites Galen about five times less frequently but Avicenna about three times more often than the *De urinis*; it further lumps together as "*Salernitani*" various authorities, such as Constantinus Africanus and Gilles de Corbeil, who were named explicitly in the *De urinis*; and it routinely refers to "*Almansor*" in its numerous quotations from Rasis, an author rarely cited in Bernard's writings and absent from his *De urinis*. These differences seem to suggest a later date of composition for the *Opusculum*, perhaps towards the middle of the fourteenth century, when most commentaries on Avicenna's *Canon* and on Rasis' *Liber ad Almansorem* were written.[98]

A feature shared by both the *Opusculum* and Bernard's *De urinis* is the importance attached to the conditions which are to be observed for accurate uroscopy because the diagnostic value of urine could be impaired by the sample's staleness, by the patient's diet, activities and emotions, or by outright deception.[99] The danger of deception, apparently a perennial challenge to the practitioner,[100] is the principal theme of another treatise ascribed to Bernard de Gordon, *De cautelis urinarum*.[101] This collection of instructions for the diagnosing physician was commonly associated with the name of Arnold of Villanova as *Cautele medicorum*, but it was

[97] For tertian fever, "quia iuvenes multum gaudent experimentis, ideo aliqua ponamus": *De conservacione*, 2: *De urinis*, in Vatican MS Pal. 1174, fol. 42v.

[98] Thus, at Montpellier around 1350, commentaries on the fourth *Canon* were written by Bernardus Alberti, Petrus Albi, Johannes Jacobi, and Gerardus de Solo; commentaries on the ninth *Liber ad Almansorem* by Gerard de Solo and Johannes de Tornamira.

[99] Bernard lists no fewer than twenty-one "condiciones, quas debemus observare in iudicio urinarum, ut nullus error intercadat": *De conservacione*, 2: *De urinis*, 5, in Vatican MS Pal. 1174, fol. 31v.

[100] We may recall, for example, the delightful tenth-century anecdote of Duke Henry of Bavaria whose attempt to substitute a maid's urine for his own was foiled by Notker Balbulus: see Loren C. MacKinney, *Early Medieval Medicine* (Baltimore, 1937), pp. 45-46.

[101] *Nota de cautellis* [sic] *necessariis medico ad urinas. Gordonius.* [*Incipit*] "Primo pete semper cuius est. Si dicit mulieris... [*desinit*] ... a subtilioribus et liquidioribus incipe. Da syrupos digestivos etc.," in Oxford MS Bodl. canon. misc. 455, fols. 237r-238v. Printed with the *Lilium medicine*, for example in the Venice 1498 and Paris 1542 editions. A different version seems to be extant in London Wellcome MS 548, 15th c., fol. 48r-v, which I have not yet examined.

actually excerpted from several writings and it circulated in varying versions.[102] The accent of the instructions is on maintaining the physician's authority and prestige in dealing with patients and their relatives, especially when they are either too anxious and inquisitive or uncooperative and disingenuous.

The strategems, proposed in the *Cautele* for the practitioner's self-defense, show a keen sense of psychology if not of morality. Their cynicism is in sharp contrast with the ethical tone of another work on the physician's conduct that was also (and, one likes to think, more justly) attributed to Bernard de Gordon, the *Tractatus de modo medendi in egritudinibus presertim cum medicus nescit causam*.[103] In the *Cautele*, for example, the physician is advised: "always do something new so that the bystanders will not deem you ignorant" or unable to do anything without books.[104] In the *Tractatus*, however, he is encouraged to study diligently in the books and even to be humble enough to seek the advice of "someone more experienced."[105] The author of the *Cautele* is not averse to the use of some trickery by the image-conscious practitioner:

> Promise health to the patient who is hanging on your lips, but once you have left him say some words to the members of the household to the effect that he is very ill; then, if he recovers you will be praised more, but if he dies his friends will testify that you had almost given him up.[106]

[102] See Henry E. Sigerist, "Bedside Manners in the Middle Ages; The Treatise *De cautelis medicorum* Attributed to Arnald of Villanova," *Quarterly Bulletin* of Northwestern University Medical School Chicago, 20 (1946), 136-143. The text translated there by Sigerist (from the Basel 1585 edition of Arnold's *Opera*) is in several ways different from that found in Munich Clm 456, 15th c., fols. 3r-6v, *De modo tenendi quem* [sic] *usus est magister Arnoldus de Villanova cum visitavit infirmos*, as well as from that attributed to Bernard de Gordon.

[103] [*Incipit*] "Cum medicus ut plurimum a principio egritudinis morbum exquisite et in sua racione speciali cognoscere minime possit. ... Explicit tractatus Bernardi de Gordonio de modo medendi servando a medico antequam cognoscit morbum," in Vatican MS Pal. 1083, fol. 322r-v.

[104] "Cum vadis ad infirmum semper fac aliquid de novo ne astantes reputent te ignorantem," in Oxford MS Bodl. canon. misc. 455, fol. 273r; also see Sigerist, "Bedside Manners," p. 140.

[105] "Laboret medicus de signis seu accidentibus iudicare, cuius [*sic*] noticiam perfecte poterit habere ex libris sciencie medicine si ipsorum diligens fuerit inquisitor. Quo facto si proprio studio non possit medicus medicamen conveniens tribuere debet studere qualiter alterius studio hoc poterit facere et tunc ad alium periciorem recurrat," in Vatican MS Pal. 1083, fol. 322r.

[106] "Egro ab ore tuo pendenti sanitatem promittas. Cum vero ab eo recesseris domesticis verbis dicas ipsum multum laborare: tunc si ipse liberatus fuerit maioribus meritis et laudandus eris. Si autem mortuus fuerit testabuntur te amici sui de illo te quasi desperasse," in Munich Clm 456, fols. 3v-4r; also see Sigerist, "Bedside Manners," p. 142.

Such double-dealing is explicitly rejected by the author of the *Tractatus*:

> If the physician, while leaving, is questioned by the patient's friends as to what he thinks about his condition, he should not answer with fraudulent or wicked duplicity, claiming to hope for what he does not anticipate or to fear what he does not fear, but he should answer with cautious discretion, honestly and within the limits of truth, and say that he is not quite sure about the patient's condition.[107]

In its emphasis on integrity, epitomized in the injunction that "the physician should visit his patient diligently,"[108] the *Tractatus* is the direct antithesis of the *Cautele*, which nonchalantly suggests that he "should not go but send a messenger" to a patient in danger.[109]

In addition to contrasting standards of professional conduct, a different objective further distinguishes these two treatises. While the *Cautele* concentrated on diagnosis, the *Tractatus*, as its full title indicates, was concerned more with the procedures the physician should follow when he was unable to diagnose adequately.[110] Until he knew the specific disease, he might prescribe "temperate" medicines that could "help in some way and do no harm," for example borage or comfrey, but with due caution for possible side-effects: "thus, you should not administer camomile to a pregnant woman because it provokes abortion."[111] The most important provisional treatment should be dietetic and should maintain a "neutral regimen" of the six non-naturals, especially food and drink; the recommended menu consists of barleymeal, almond milk, lemon juice and – not surprisingly – chicken broth.[112]

[107] "Si dum medicus recedit ab amicis egri interrogetur quid sentenciat de statu infirmi respondere debet non cum duplicitate fraudulenta vel maliciosa scilicet de deo [?] se sperare quod non intendit aut se timere quod non timet sed debet respondere cum cautela discrecionis honeste vel lima veritatis ut dicat quod non bene est certus de statu infirmi," in Vatican MS Pal. 1083, fol. 322r.

[108] "Medicus debet suum infirmum diligenter visitare" (ibid.).

[109] "Item si est in periculo non vadas ad eum sed mitte nuncium": in Oxford MS Bodl. canon. misc. 455, fol. 237r.

[110] The same concern led Arnold of Villanova to compose his work that is printed as *Tabule que medicum informant specialiter cum ignoratur egritudo*, in *Opera* (Lyon, 1509), fols. 118v-119v. The *Tabule* omit moral implications, and their systematic arrangement must have made them more useful than the *Tractatus*.

[111] "[Medicus] ordinet aliqua medicinalia que aliquo modo possunt proficere et nullo modo nocere. ... Cave tamen ne temperatam medicinam habentem utilem proprietatem et aliam inutilem non [*sic*] administres quemadmodum camomillum mulieri pregnanti administrare non debes quia procurat aborsum," in Vatican MS Pal. 1083, fol. 322r-v.

[112] "Ideo ipsi egro regimen neutrum debet administrare. ... Regimen in dieta consistit in commensuracione rerum non naturalium. ... In principio enim administretur ordeatum, in augmento lac amigdalis dulcis, in statu aqua citreorum, in declinacione brodium pullarum" (ibid.).

Therapeutic procedure is the subject of two more treatises found under the name of Bernard de Gordon. The first, called *Modus medendi extractus de modo Coffonis*,[113] is a digest of the *Ars medendi* of the Salernitan physician Copho (d. 1110).[114] Although they do not necessarily preclude Bernard's authorship of this digest, some substantial differences with his work *De decem ingeniis curandorum morborum* are worth noting. In Bernard's work, which was drawn primarily from Galen's *De ingenio sanitatis*, ten methods of treatment were distinguished and further subdivided on the basis of theoretical considerations, including the "essence of the disease," the body's complexion, and the anatomy and physiology of the organs.[115] The *Modus medendi*, on the other hand, simply divided therapy into four practical approaches according to the respective category (purgative, constrictive, restorative, and digestive) of medicines used. Still another criterion is applied in the *Modus practicandi* "according to Bernard de Gordon," which offers a "brief introduction to medical treatment (*medendi breves ysagogas*)" at the request of a young friend. This introduction distinguishes only two therapeutic modes, one for summer and the other for winter, and describes them metaphorically as two "swords with which to strike down and cut to pieces the phalanges of disease."[116]

While it seems plausible that this note on treatment by seasonal diet and pharmacy is a genuine record of some impromptu advice by Bernard, it is doubtful whether he was in any way responsible for a few other works on medication that also occur under his name in manuscripts.[117] Their

[113] The only extant copy is acephalous, [*incipit*] "vel paucitate medicine cui adhibeatur remedium Ypocratis dicentis, Si vis ducere ellectuarium...", and *desinit*, "Fit eciam alteracio tam in cibis quam in potibus quam in medicinis convenientibus etc. Explicit modus medendi magistri bernhardi de Gordonio extractus de modo Coffonis," in Munich Clm 19901, mid-15th c., fols. 1r-11v (rather than -10, as in Sudhoff, "Schriftstellerei," p. 182).

[114] Copho, *Ars medendi*, ed. de Renzi, *Collectio Salernitana*, 4 (Naples, 1859), pp. 415-438.

[115] See chapter 2, pp. 47-48 above.

[116] *Modus practicandi secundum magistrum Bernardum de Gordonio* [this title is in a different, slightly later hand]; [*incipit*] "Ego fili carissime tuis acquiescens precibus medendi breves ysagogas tibi tradere non negligo. In primis itaque tibi trado duos gladios unum estivalem alium hyemalem cum quibus egritudinis falanges confodias et trucides. ... [Last ten lines of the text are nearly illegible because of water damage.] Explicit modus practicandi secundum Bernardum de Gordonio," in London British Library MS Sloane 134, 15th c., fols. 170r-172r. Also, according to TK 487, in Oxford Oriel College MS 4, 14th c., fols. 229-(230), which I have not yet examined.

[117] Printed with the *Lilium medicine* (Lyon, 1574), pp. 948-963, is a "Bernardi

attribution to Bernard may have been based merely on the reputation of his *Lilium medicine* and especially of the *Antidotarius* appended to it.[118] In addition, most of those works deal with materia medica, a subject that usually provides few clues to the compiler's identity,[119] and several of them are found elsewhere attributed to other authors. Thus, a treatise *De conferentibus et nocentibus*, which lists the beneficial and harmful medicinal substances for each part of the body from the head on down, was ascribed not only to Bernard[120] but also to Galen, Johannes Platearius, Gualterus Agulinus, and Arnold of Villanova,[121] and it is very similar to the *Areolae* of Johannes de Sancto Amando (fl. 1300).[122] Bernard's alleged authorship of a *Compendium compositionum* may seem confirmed in one instance where this brief tract on compounds is inserted between two authentically Gordonian treatises in a relatively early codex,[123] but it can not be ascertained without disentangling the complex textual relationship of this compendium with many similar works,[124] primarily with the ubiquitous *Practicella ex dictis Mesue abbreviata* by Johannes de Parma (fl. 1250–1270),[125] but also with those attributed to

---

Gordonii de Floribus dietarum liber unus," [*incipit*] "Corpus hominis ex quatuor constat humoribus. ..." However, no author is named in four Erfurt Amplonian manuscripts (F.237, F.275, F.289, and Q.185) that contain this work *De floribus dietarum*, and the Erfurt catalogue attributes it to Arnold of Villanova: see Sudhoff, "Schriftstellerei," p. 181.

[118] The *Antidotarius* occurs as a separate work not only in Latin (for example in Paris BN MS lat. 6966, a. 1461, fols. 1r-3r; and in London BL MS Sloane 2527, 15th c., fols. 146v-153v), but also in vernacular versions (thus, in French, in Paris BN MS fr. 1288, 15th c., fols. 136-142; and in Provençal, in Paris Arsenal MS 2525, item 8).

[119] I have not yet been able to verify the tentative attribution in TK 352, "(Bernard Gordon?)," of a treatise *De medicinis conficiendis*, [*Incipit*] "Cum volueris conficere medicinam prius consideranda sunt...," extant in three early manuscripts.

[120] [*Incipit*] "Conferunt cerebro fetida ut in gravi...," in Oxford MS Bodl. canon. misc. 455, 15th c., fols. 226r-237r. Also in Cambridge University Peterhouse College MS 178, 13th c. [?], fols. 202v-207v. Also perhaps in Oxford Oriel College MS 4, 14th c., fols. 226-(229): see TK 914 (with question mark).

[121] See TK 246.

[122] Johannes de Sancto Amando, *Areolae de conferentibus et nocentibus*, ed. Julius Pagel (Berlin, 1894).

[123] [*Incipit*] "Quoniam de melioribus amicis quos habere videor...," in Cambridge Gonville and Caius MS 373 (593), 14th c., fols. 101r-102v; where it is preceded by Bernard's *Regimen acutorum* and followed by his *Tractatus de gradibus*.

[124] Including one labelled *Gordonius de dosi medicinarum* in Erfurt Amplonian MS F.236, fols. 94r-95v: though inserted between Bernard's *De pulsibus* and his *De pronosticis*, this is actually the work of Johannes de Parma (see n. 125 below) according to Sudhoff, "Schriftstellerei," pp. 181-182.

[125] For manuscripts and references see TK 1295.

several other authors,[126] including one intriguing "Bernardus de Villanova."[127]

Bernard de Gordon and Arnold of Villanova, who shared credits for such treatises as *De sterilitate*, *De floribus dietarum* and *De conferentibus et nocentibus*,[128] may occasionally have been remembered as interchangeable or in close association on account of their frequently overlapping interests. In other instances it was the contrast between their personalities, of which we have noted some indications earlier,[129] that was underscored when to each was attributed a different work on the same subject. This is clearly the case for two treatises on astrological seals and images, as may be seen from the following illustrative excerpts and brief comparative analysis. According to the *Tractatus magistri Bernhardi de Gordonio ad faciendum sigilla et ymagines contra infirmitates diversas*,[130]

> Libra is a masculine air sign whose figure should be that of a man having a scale in his hands. It aids against disease of the uterus, spleen, and stomach. The imprint should be made of a man who has twenty [*sic*] hands and holds in his right hands [*sic*] a scale and in his left hands [*sic*] a book, or vice versa. It may be made of any metal, when the sun is in the first face of Libra; Venus should not be retrograde and the moon should be crescent, and if it is in Libra it will be very good. It should be made only on the day and at the hour of the sun, and only by striking. Or the imprint may be made according to what Enoch said: the figure of a man having in his hands two straight scales and at the tip of one hand a ship that is half white and half black; of any metal, as has been said, weighing seven grains of wheat, and when the sun is in the first face of Libra and Venus is not retrograde nor Jupiter in Libra. It may also be made on Thursday, between the first and the fifth hour. The *deodii* [?] sages, however, say that it should be made only on the day and at the hour of the sun. And Enoch said that if the sun should strike the figure directly it would be very good. There are many considerations about this figure, but these are the necessary and most useful ones on which all agree. An experimenter who made one,

[126] For example, Antonius de Parma (TK 1270), John of Gaddesden (TKa 101), John of Montpellier (TK 1295), and Marsilius de Sancta Sophia (TK 1244).

[127] In a *Practica* in two fifteenth-century Cambridge manuscripts: see TK 1270.

[128] See notes 67, 117, 120 and 121 above.

[129] See pp. 35-36 above.

[130] [*Incipit*] "Incipiamus cum Dei adiutorio de hijs que possunt fieri sole existente in quolibet signo ... Aries est signum masculinum igneum...," in Vienna Nationalbibliothek MS 3162, 15th c., fols. 239r-241r. Also, as *De imaginibus signorum*, [*incipit*] "Aries est signum femineum [*sic*] igneum...," in Wiesbaden MS 79, a. 1518-1520, fols. 54v-56r. I hope to publish a critical edition and more detailed analysis of these texts in the near future.

furthermore, said that it should be made at the beginning of the abovesaid fifth day when Jupiter is above the earth, and it should not be struck but painted; and he found it marvelous. He was also of the opinion that the cause of its efficacy was the fact that Jupiter was above the earth.[131]

The corresponding paragraph in the *Sigilla magistri Arnaldi*[132] is quite different:

One should take the purest gold, melt it, and shape a round seal while saying, "Lord, arise to my aid and hear my voice because I have called out to you: have mercy on me and hear me"; and say the psalm, "The Lord is my light etc." This should be done while the sun is entering Libra, that is on or after the fifteenth kalends of October, while the moon is in Capricorn or in Aquarius. Engrave on one side the figure of a man who in his hands holds a scale in the fashion of a cross, with salt on the scale; and around him [the words], "Hely hely lamazabathani: it is fullfilled"; and on the other side in the circumference, "Jesus the Nazarene, King of the Jews," with in the center Michael, iota, Matthew, vau. This most sacred seal works against the snares of demons on earth and sea, and it delivers the bearer from sudden death. Who bears it will be meek, merciful, wise, honest, and useful in giving counsel. It also helps in acquiring profit in any commerce, and it helps in the devoted love of both men and women; and who bears it with the seal of Aries will not be unjustly accused before the sovereign. It also helps against the diseases coming from the blood, against the forces of winds and the floods of seas, and against kidney ailments. Evil spells and witchcraft will not harm the house where it is, and it is helpful for many

[131] "Libra est signuo aereum masculinum cuius figura sit homo habens libram in manibus suis. Et subvenit egritudini matricis, splenis et stomachi; fiat ergo impressio hominis habentis 20 [*sic*] manus, tenentis in manibus dextris libram et in sinistris librum vel e converso. Fiat ex quocumque metallo sole in Libra existente in prima facie ipsius et non sit Venus retrograda et luna sit in augmento, et si sit in Libra erit valde bonum; fiat quoque in die et hora solis tantum et non aliter quam percutiendo. Vel aliter fiat impressio secundum quod dixit Enoch figura hominis habentis in manibus duas libras rectas et in capite unius manus navem unam mediam albam et mediam nigram, ex quocumque metallo sicut dictum est ad pondus 7 granorum frumenti sole existente in prima facie Libre et non sit Venus retrograda nec Jupiter in Libra. Fiat eciam in die Jovis a prima hora usque ad 5. Dicunt tamen sapientes deodii quod debeat fieri in die et hora solis tantum. Et dixit Enoch quod si sol percuteret figuram directe esset valde bonum. Multe sunt condiciones istius figure, sed hec sunt necessarie et magis utiles in quibus omnes conveniunt. Experimentator quoque qui fecit eam dixit quod debet fieri in principio diei quinti ante dicti Jove existente super terram, non percutiendo sed pingendo et invenit eam mirabilem. Et opinatus est quod causa efficacie istius sit quod Jupiter erat super terram," in Vienna MS 3162, fol. 240r-v.

[132] In Paris MS Bibliothèque Nationale lat. 7468, s.d., fols. 39-40v (TK 1320), and, probably also in BN lat. 7349, 15th c., fols. 127r-129v (TK 26), as well as in the printed editions of Arnold's works.

> other things. Who carries it with him will safely sail the sea, always imploring God's mercy and asking His forgiveness for sins. One should carry it with reverence and with fear of God.[133]

Although both treatises fall entirely within the realm of magic, the *Tractatus* is pragmatic and down to earth like its presumed author Bernard, while the *Sigilla* is more visionary and mystical, like Arnold.[134] In style and logical arrangement the *Tractatus* surpasses its wordy and somewhat confused counterpart and reminds us of Bernard's own works and of his "translation" of the *Reductio geomancie ad orbem*.[135] Its use of sources heightens the contrast with the *Sigilla*, which contains no references: throughout the *Tractatus* the hearsay character of its information is acknowledged repeatedly and diverse viewpoints are presented. It may be noted that this comparative presentation, as well as the qualifying phrase "some say (*dicunt quidam*)," and even such citations as *sapientes indi*, *sapientes egipcii* and *experimentator* are also found in Bernard's *Lilium medicine*.[136] A further difference between our two astrological texts lies in the nature of their instructions for engraving the

[133] "Accipiatur aurum purissimum et fundatur; post formetur sigillum rotundum et dum formabitur dicas, Exurge domine in statere et exaudi vocem meam, quia clamavi ad te; miserere mei et exaudi me. Et dicas psalmum, Dominus illuminatio mea etc. quod cum factum fuerit reponatur; et hoc fiat sole intrante in Libra, videlicet xv kalendis octobris et post luna existente in capricorno vel in aquario. Sculpatur ab una parte figura hominis in manibus tenens libram ad modum crucis, sale existente in libra et in circumferentia sui sit hely hely lamazabathani, consummatum est; et ab alia parte in circumferentia, Jesus nazarenus rex iudeorum; et in medio Michael, Joth, Mattheus, Vau. Valet autem istud sigillum sacratissimum contra insidias demonum in terra et in mari, et a morte subitanea liberat portantem se, et qui portat eum erit mansuetus et misericors, sapiens et honestus et ad dandum consilia utilis. Valet etiam ad lucrum acquirendum in mercaturis quibuslibet, et valet ad dilectionem tam virorum quam mulierum bono zelo; et qui eum portat cum sigillo arietis iniuste non poterit coram principe accusari; et valet contra infirmitates que fiunt ex sanguine et contra impetus ventorum et marium inundationes, et contra dolorem renum, et in domo ubi fuerit maleficia et sortilegia illi domui non nocebunt; et valet ad multa alia. Et qui eum portabit secum securus per mare navigabit implorando semper dei misericordiam, et ab eo de peccatis veniam postulantem portetur autem cum dei reverentia et timore," in Arnold of Villanova, *Opera* (Lyon, 1509), fols. 301v-302r.

[134] For an interesting illumination of this contrast between both personalities, see Michael McVaugh, "Nota sobre las relaciones entre dos maestros de Montpellier: Arnau de Vilanova y Bernardo Gordon," *Asclepio, Archivo iberoamericano de historia de la medicina y antropologia medica*, 25 (1973), 331-336.

[135] See pp. 43-46 above.

[136] "Indi" and "Egipcii": *Lilium medicine*, in London BL MS Harley 3698, fols. 21r and 24r; also in *Liber pronosticorum*, in Vatican MS Pal. 1174, fol. 100r. "Experimentator": *Lilium medicine*, fols. 11r and 68v. The other authors cited in the *Tractatus* are Enoch (for whom see Thorndike, *Magic*, vol. 1, chapter xiii and passim) and "sapientes deodii," whose identity I have thus far been unable to determine.

seals. The *Tractatus* focuses on the ideal astrological times and on the iconography of the zodiacal signs, occasionally adding peculiar details, for example in the description of Gemini as "a man holding by the hand a woman who lacks nails on her hands and feet; the man's face is round, that of the woman is oblong."[137] The *Sigilla*, on the other hand, emphasizes the required incantations and the textual inscriptions, while also assigning the figure of an apostle to the reverse of each seal.[138] The gnostic vein of these formulae and figures, which was pointed out by Lynn Thorndike,[139] is complemented by the supernatural efficacy attributed to several seals against a wide range of ills. The talismans are guaranteed to ward off not only diseases but also natural disasters, demons and witchcraft; and the bearer is warned to abstain from "impurity, lust and other mortal sins."[140] This supernatural orientation is absent from the *Tractatus*, where the efficacy of the seals is restricted to medical problems and its only suggested explanation is based on astrological conjunctions.

It should be made clear that the purpose of contrasting the two treatises on seals was neither to discredit nor even to characterize Arnold's versatile scholarship but rather to underscore Bernard's image as projected in the attribution to him of certain writings. Among those writings the *Tractatus ad faciendum sigilla*, with its "superstition" essentially limited to the premise of astrological efficacy, is the only one that deals with the supernatural. Among the works ascribed to Arnold, on the contrary, a considerable portion is comprised of treatises on magic and alchemy.[141] If such treatises needed a distinguished name to boost their respectability, Arnold seemed a logical choice because of his association with popes and kings, his penchant for speculative theory, and his excursions into

[137] "Gemini est signum masculinum rubeum habens 2 corpora virum tenentem mulierem in manu que caret ungulis in manibus et pedibus et facies hominis est rotunda mulieris vero longa," in Vienna MS 3162, fol. 239r. Other peculiar details occur in the descriptions of Cancer ("fit sine oculis et naso": ibid., fol. 239v), Leo ("fit sine virga, testiculis et auribus": ibid.), Scorpio ("non habens oculos nec testiculos ad faciendum malum": ibid., fol. 240v), and so on.

[138] The individual apostles assigned to each sign of the zodiac in the *Sigilla* appear different from those in the general astrological tradition.

[139] Thorndike, *Magic*, 2: 858. Professor Thorndike discussed similar gnostic talismans of an earlier era in *Magic*, vol. 1, chapter 15, especially pp. 361-368.

[140] "Et qui portat eum [*sic*] caveat inquantum poterit ab immunditia et luxuria, et ab aliis peccatis mortalibus," in Arnold of Villanova, *Opera*, fol. 301v.

[141] The most recent and reliable bibliography of Arnold's authentic and spurious works is by Juan Antonio Paniagua, *Estudios y Notas sobre Arnau de Vilanova* (Madrid, 1963).

mysticism. Bernard apparently lacked such credentials, and the Gordonian writings discussed above, whether authentic or spurious, are consistent with his more modest stature. His purview was restricted to the province of medicine, his theorizing was rather sober and always tied to practical applications, and practice was his principal concern. This characterization of Bernard of Gordon as professor and practitioner is not only documented in the form and subject matter of the writings associated with his name, but it is also illuminated in the substance of his thought which will be analyzed in the following chapters.

4

Mainsprings of Bernard's Teaching: Tradition, Reason, Experience, Nature

Bernard's writings, examined separately in the preceding chapters, may also be considered together as an organic *corpus* in order to determine the mainsprings and characteristics of his scholarship. The most prominent feature appears to be the combination of traditional knowledge with his own personal views. This feature is hardly accidental, for the combination of bookish and personal knowledge heads the requirements which Bernard himself deemed essential for good teaching. In his *Regimen sanitatis* he asserted that, above all, the master should combine "subtlety and ingenuity" in interpreting the sources and in selecting "the better ones" among them. Also, where the authorities (*doctores*) appeared to contradict each other, the master ought "to know how to reconcile them."[1] This already implied that even a so-called compilation, such as Bernard's *Lilium medicine*, would not be merely a matter of copying but would also require both the author's discretion in selecting the pertinent sources and his dialectical skill in settling real or seeming contradictions. It is worth observing here that Bernard directed his reconciling efforts primarily to the contradictions within the medical tradition, while accepting such conflicts of opinion as those between physicians and philosophers.[2] He thereby differed from his contemporary at Padua, the "Conciliator" Peter of Abano (ca. 1250–1316), whom he may have met at Montpellier around 1306.[3]

[1] "Prima condicio que requiritur ad hoc quod magister sit aptus ad docendum quod sit subtilis et ingeniosus ut sciat interpretari scripturas, et ubi videntur contrarietates doctorum quod sciat ad concordiam reducere et ex scripturis sciat eligere meliora": *De conservacione*, 4: *Regimen sanitatis*, in Vatican MS Pal. 1174, fol. 58v.

[2] See chapter 3, pp. 78-81 above.

[3] The most recent analysis of Peter of Abano's *Conciliator differentiarum philosophorum et praecipue medicorum* is in the study of Professor Nancy G. Siraisi, *Arts and Sciences at Padua: The* Studium *of Padua before 1350* (Toronto, 1973), especially pp. 152-

The mere ability to select and to reconcile sources was only a prerequisite to good teaching according to Bernard. Explicitly emphasizing the importance of the master's more personal contribution, he remarked that "one who has learned nothing for himself will not teach anything of significance." He further elaborated this theme in his assertion that "those who say only what they have found in booklets are no teachers (*doctores*) but are like rumormongers, because it is a poor method to use always what is found [elsewhere]."[4] At the same time, however, he insisted that the master should remain humble when propounding his own views. This apparently was not always the case, since Bernard continued, "some, when lecturing, claim that they know everything and that others know nothing."[5] From humility and the general moral requirement of exemplary conduct,[6] he then returned to the teacher's academic qualifications. The master should be eloquent, "because knowledge without eloquence is like a sword in the hand of a paralytic."[7] Bernard concluded the list of requirements with "teaching skill," which he defined as intelligibility: "one sure sign of a knowing man is that everyone understands him when he lectures."[8]

Although master Bernard formulated these standards late in his career, namely in 1308, he appears to have lived by them quite consistently. In the first place, he displayed a broad knowledge of the available sources; secondly, he selected those authorities which he found most convincing, and he frequently endeavored to solve confusing contradictions; furthermore, he regularly added his own views, even when these might not agree with generally accepted traditions, but he also humbly acknowledged the

---

153 with regard to the relation between philosophy and medicine. For Peter's possible passing through Montpellier, see Sante Ferrari, "Per la biografia e per gli scritti di Pietro d'Abano," *Memorie delle Accademia dei Lincei, Classe di scienze morali, storiche e filologiche*, series 5, 15 (1918), 649.

[4] "Nil enim magnificum docebit qui a se nichil didicit, et hoc est quod scribitur libro sapiencie. Illum eruditorem elige quem magis invenis in suis quam in alienis. Qui enim non dicunt nisi quod in cartabellis inveniunt non sunt doctores set sunt sicut illi qui narrant rumores, quoniam miserum ingenium est semper inventis uti": *De conservacione*, 4: *Regimen sanitatis*, in Vatican MS Pal. 1174, fol. 58v.

[5] "Tercia condicio quod in docendo [magister] sit humilis non superbus. Aliqui enim cum legunt dicunt se omnia scire et alios ignorare, et ita iactant se ipsos laudantes et alios vituperantes" (ibid.).

[6] "Quod sit bone vite et honeste conversacionis..." (ibid.).

[7] "Quod sit eloquens, quoniam sciencia sine eloquencia est sicut gladius in manu paralitici..." (ibid.).

[8] "Quinta condicio que requiritur est pericia docendi; unum certum signum hominis scientis est quando omnes cum legit intelligunt eum...": ibid., fols. 58v-59r.

limits of his own knowledge; and finally, he aimed for eloquence and clarity in his teaching, with remarkable success if we may judge from the style and organization of his writings.

The wide range of Bernard's medical authorities becomes evident in the following table (Table 1) of explicit references in his works.[9] To be sure, this table should not be read as the exact measure either of Bernard's knowledge of the sources or of his dependence on tradition. On the one hand, of the authors whom he named, he may have known several only indirectly, from commentaries and compendia. This would seem to be the case for such "minor" authors as Alkindi and Serapion. Conversely, authors whom Bernard did not name were undoubtedly the source of several views which he presented as his own or, more accurately, as commonly known. These views included not only medical theories but also philosophical ideas, astrological beliefs, botanical lore, and so on.[10] It can also safely be assumed that his dependence upon such standard sources as Dioscorides' *De materia medica* or Joannitius' *Isagoge in Artem parvam Galeni* was far greater than the table would indicate. Furthermore, not reflected in this table are such numerous but vague allusions as to "the ancient philosophers," to "the Indian sages," or to "our masters."

The percentage column of Table 1 reveals the proportions among Bernard's explicit references. Ideally and for the sake of perspective these proportions should be compared with the citations in the entire opus of several other authors, contemporary and non-contemporary, of Montpellier and elsewhere. At this time it is possible only to give a limited idea by adding a second table (Table 2) with the percentages and grand totals of Bernard's opus and his most substantial compositions juxtaposed to two books with a comparable size and scope, Gulielmus de Saliceto's *Summa conservationis* (ca. 1280) and John of Gaddesden's *Practica medicine* (ca. 1314),[11] and to two works whose references have been conveniently

[9] Not included in this table are most of the authorities pertaining to the liberal arts, for which see chapter 1 above. Explicit references to those authorities (including the Bible) number slightly over one hundred.

[10] With regard to astrological lore, Professor Richard Lemay has brought to my attention several instances in which Bernard apparently drew his information from the pseudo-Ptolemaic *Centiloquium* without acknowledging any source.

[11] The computations are based on the following incunabula: Gulielmus de Saliceto, *Summa conservationis et curationis* (Venice, 1489) and John of Gaddesden, *Rosa anglica practica medicine* (Pavia, 1492), both at the New York Academy of Medicine. For Gaddesden see also above, chapter 2, p. 58 and note 118.

TABLE 1. BERNARD'S MEDICAL AUTHORITIES

Sources	*Regimen acut.* (1294)	*Compendium reg.* (1294)	*Pronost.* (1295)	*Ingenio* (1299)	*Gradibus* (1303)	*Lilium medicine* (1303)	*Conserv. vite* (1308)	1. *Flebotomia*	2. *Urinis*	3. *Pulsibus*	4. *Regimen sanitatis*	*Tyriaca* (n.d.)	*Marasmo* (n.d.)	Total	Percentage
Greek	5	4	7	30	19	317		111	109	7	149	16	16	790	64.8
Hippocrates	3	3	4	3	1	80		7	14	–	20	1	4	140	11.4
Aristotle	–	–	–	–	–	8		1	3	1	19	4	4	40	3.2
Dioscorides	–	–	–	–	–	3		–	–	–	–	1	–	4	
Galen	2	1	3	27	18	226		103	86	6	110	10	8	600	49.2
Judeo-Arabic and Salernitan	2	–	–	6	9	164		81	46	6	69	15	16	414	33.7
Alkindi	–	–	–	–	2	–		–	–	–	–	–	–	2	
Joannitius	–	–	–	–	–	1		–	–	–	–	–	–	1	
Serapion	–	–	–	–	–	2		–	–	–	–	–	–	2	
Damascenus	–	–	–	–	–	5		–	1	–	–	–	–	6	
Rasis	–	–	–	–	–	1		9	–	–	–	–	–	10	
Isaac	–	–	–	–	1	1		–	15	–	2	–	–	19	
Haly [Abbas]	–	–	–	1	2	4		12	1	2	4	–	–	26	
Haly [Ridwan]	–	–	–	–	–	1		5	–	1	–	–	3	10	
Heben Mesue	–	–	–	–	1	3		–	2	–	–	–	–	6	
Avicenna	1	–	–	5	1	130		55	26	3	60	8	8	297	24.3
Constantinus	–	–	–	–	–	12		–	1	–	2	–	–	15	
Nicolaus	–	–	–	–	–	–		–	–	–	–	1	–	1	
Averroës	1	–	–	–	2	4		–	–	–	1	6	5	19	
Contemporary	–	–	–	–	2	5		–	8	2	–	–	–	17	1.4
Aegidius	–	–	–	–	–	–		–	7	1	–	–	–	8	
Guiraldus	–	–	–	–	–	2		–	–	–	–	–	–	2	
Alveredus	–	–	–	–	–	–		–	–	1	–	–	–	1	
[Barthol. Anglicus]	–	–	–	–	–	–		–	1	–	–	–	–	1	
Albertus Mg.	–	–	–	–	–	2		–	–	–	–	–	–	2	
Anselmus Jan.	–	–	–	–	–	1		–	–	–	–	–	–	1	
Arnold Vill.	–	–	–	–	2	–		–	–	–	–	–	–	2	
Total	7	4	7	36	30	486		192	163	15	218	31	32	1221	

tabulated in earlier studies, Henri de Mondeville's *Cyrurgia* (1306) and Guy de Chauliac's *Magna Chirurgia* (1365).[12]

TABLE 2. BERNARD AND HIS CONTEMPORARIES
(Percentages and Grand Totals)

Sources	Bernard's opus	*Lilium* (1303)	*Cons. vite* (1308)	Saliceto *Summa cons.* (ca. 1280)	Gaddesden *Practica* (ca. 1314)	Mondeville *Cyrurgia* (1308)	Chauliac *Chirurgia* (1365)
Greek	64.8	65.2	63.9	44.0	37.0	42.0	33.8
Hippocrates	11.4	16.0	7.0	28.6	7.9	5.2	4.0
Galen	49.2	46.5	51.8	10.6	27.5	33.0	27.0
Judeo-Arabic and Salernitan	33.7	33.0	34.3	55.0	57.0	33.5	42.5
Rasis	0.9	–	1.5	16.0	1.6	5.0	5.0
Avicenna	24.3	26.7	24.4	35.0	31.2	23.6	20.0
Contemporary	1.4	1.0	1.9	1.0	5.0	24.5	23.7
Totals:							
—authors cited	24	18	16	8	46	58	100
—single citations	1221	486	588	272	1525	1308	3300

The most conspicuous feature of Table 2 is the rise in total numbers of citations, which is probably due not only to the growing volume of the later treatises but also to the natural accumulation of available information and to the expansion of the working libraries at the compilers' disposition. Further study would be needed, including a computation of the ratio between the citations and the length of each work, to determine whether scholastic authors made increasingly systematic efforts to identify their sources.[13]

[12] The percentages were computed from data supplied in Sarton, *Introduction*, 3: 870 for Henri de Mondeville's *Cyrurgia*, and in *La Grande Chirurgie de Guy de Chauliac*, ed. E. Nicaise (Paris, 1890), p. xxxv.

[13] As I assumed and tentatively documented in my article, "Scholasticism in Compendia of Practical Medicine, 1250-1450," *Manuscripta*, 20 (1976), 87-88.

The other features of this comparative table may serve us as touchstones in the examination of Bernard's explicit references. With regard to the proportions between the different strata of tradition, is it especially noteworthy that two thirds of his references are to Greek authors, especially Hippocrates and Galen, against one third to Judeo-Arabic authors, such as Avicenna, Haly Abbas and Isaac Judeus, and to Salernitan sources, namely Constantinus Africanus and Nicolaus Salernitanus. The scarcity of "contemporary," that is to say thirteenth-century authors among those identified by Bernard seems most peculiar.

In spite of the fact that he rarely cited recent sources there are several indications that Bernard regularly consulted not only general encyclopedias, especially the *De proprietatibus rerum* of Bartholomeus Anglicus,[14] but also medical compendia such as that of Gilbertus Anglicus.[15] Such works were probably too commonly known to require identification. A similar explanation might apply to Bernard's vague allusion to an unnamed *versificator* as the author of mnemonic verses, including those taken from the *Regimen sanitatis Salernitanum*[16] or from the *Flos medicinae Scholae Salerni*. A rhyme from the latter has indeed become famous through Bernard's quotation,[17] which deserves to be cited here:

> The infectious diseases are contained in this verse: Acute fever, phthisis, epilepsy, scab, erysipelas, anthrax, lupus, and leprosy are known as contagious.[18]

[14] See above, chapter 1, p. 17 and notes 80-82.

[15] Thus, for example, Bernard's discussion of leprosy (*Lilium medicine*, 2.25) is in several respects similar to that of Gilbertus Anglicus (*Compendium medicine*, 2): it is, of course, quite possible that both used the same source.

[16] For example, "De regimine autem sanitatis versificator aliqua dixit sic, Si vis incolumem, si vis te reddere sanum, curas tolle graves, irasci crede prophanum. Parce mero, cenare cave, nec sit tibi vanum pergere post epulas, sompnum fuge meridianum. Non teneas minctum nec cogas fortiter anum," *Lilium medicine*, in London BL MS Harl. 3698, fol. 66v. See the English translation by Pascal Parente, *Regimen sanitatis Salernitanum* (New York, 1967), p. 43. As an illustration of historical continuity it may be pointed out that there exists a still popular Flemish rhyme that is a translation of the Salernitan rule of health cited here. Another example: "versificator advertens ad consuetudinem ... dixit, Cena brevis vel cena levis fit raro molesta. Magna nocet, medicina docet, res est manifesta": *Lilium medicine*, fol. 66r. See tr. Parente, *Regimen sanitatis Salernitanum*, p. 44.

[17] The *incipit* of that quotation is listed in TK 550. Also see Sarton, *Introduction*, 3: 874, and Garrison, *History of Medicine*, p. 186. A brief analysis with references is found in Julius Pagel and Karl Sudhoff, *Kurzes Handbuch der Geschichte der Medizin*, 4th ed. (Berlin, 1922), pp. 206-207.

[18] "Morbi autem infectivi in hiis versibus continentur: Febris acuta, ptisis, pedicon, scabies, sacer ignis, fluxungia [*sic*, for antrax] lippa, lepra nobis contagia prestant," *Lilium*

The difference between this verse and that in the *Flos medicinae*, which also included *frenesis* and added three kinds of eye infections,[19] indicates some of the problems in determining how faithfully Bernard depended upon tradition or how accurately he represented it. Thus, it is impossible to know whether his source was defective or altogether different,[20] or whether he quoted from faulty memory or knowingly omitted four diseases because he did not consider them contagious. Furthermore, since the number of allegedly infectious diseases ranged from twelve in the *Flos medicinae* to only five in other sources,[21] it seems unwarranted to generalize from Bernard's quotation that "in the Middle Ages it was customary to regard eight diseases as contagious."[22] An additional problem, illustrated by the verse above, relates to the precise characterization of Bernard's sources: since the association of both the *Regimen sanitatis* and the *Flos medicinae* with the "Schola Salerni" is doubtful, his quotations from those verse collections can have little value as arguments for his dependence on the Salernitan tradition.

In view of the suggestion that Bernard represented the Salernitan tradition,[23] it is further worth observing that he appears to have drawn relatively little on the writings of the School of Salerno in general and of its protagonist Constantinus Africanus (ca. 1015–1087) in particular. Although Table 1 lists fifteen references to Constantinus, Bernard actually named him only seven times – without citing a specific work:[24] in the other half of the citations he referred to the *Viaticum*[25] and to an author he called "Viaticus."[26] It is difficult to ascertain who was meant by this

---

medicine, in London BL MS Harl. 3698, fol. 38v. The list occurs earlier (with *antrax* for *fluxungia*, and *lupa* for *lippa*) ibid., fol. 16r, in the chapter on leprosy, where it also appears that Bernard equated contagious and hereditary diseases. Interestingly, the printed versions have substituted *pediculi* (lice) for *pedicon* (epilepsy): thus the Venice 1498 edition of the *Lilium medicine*, fol. 16v.

[19] *Flos medicinae Scholae Salerni*, vv. 2483-2485, ed. Salvatore de Renzi, *Collectio Salernitana*, 5 (Naples, 1859), p. 71.

[20] For example, Bernard may also be drawing on the pseudo-Aristotelian *Problemata*.

[21] See Karl Sudhoff, "Fünf ansteckende Krankheiten," *Mitteilungen zur Geschichte der Medizin und der Naturwissenschafte*, 14 (1915), 405-406.

[22] Garrison, *History of Medicine*, p. 186.

[23] Thus, Sarton, *Introduction*, 3: 873.

[24] In at least three instances (*Lilium medicine*, fols. 83v and 89v, and *De conservacione*, 4: *Regimen sanitatis*, fol. 68v) it may be inferred from the context that Bernard is referring to Constantinus' *De coitu*, for which see Constantinus Africanus, *Opera* (Basel, 1536), fols. 299-307.

[25] See chapter 3, p. 84 above, n. 64.

[26] For example, "dicebat Viaticus qui primum vinum de vite mollitus est inter sapientissimos debet reputari": *Lilium medicine*, fol. 31v. Also ibid., fols. 31r, 36v, 48v and 68v.

"Viaticus," since both Isaac Judeus (fl. 930) and his disciple Ibn al-Jazzār (d. ca. 984) share the title to the authorship of a *Guide for Physicians*,[27] which Constantinus translated as *Viaticum*,[28] and on which both Gerard of Cremona (ca. 1114–1187) and Gerard of Béziers (thirteenth century) wrote commentaries.[29] Bernard's use of the derivative epithet "Viaticus" may indicate his awareness of the compound authorship of a work such as the *Viaticum*, and of Constantinus' secondary role as a translator.[30] A more direct indication of that awareness relates to the most important work that went under the name of Constantinus, the *Pantegni*,[31] which was a free translation of the *al-Kitāb al-Malakī* of the Persian physician Haly Abbas (al-Majūsī, d. 994).[32] Bernard usually cited "Haly in the *Liber regum*,"[33] thereby apparently referring to the later and more accurate translation (*Liber regalis*) by Stephen of Antioch (1127),[34] and in one case he simply added "and Constantinus *ibidem*."[35] It would seem, therefore, that he regarded Constantinus less as an authority in his own right than as one of the interpreters of the Greek and Judeo-Arabic traditions.

The Judeo-Arabic sources themselves comprise about one third of Bernard's citations, which is considerably less than the proportion in the

[27] Isaac Judeus, *Musar Harofim*, tr. Saul Jarcho, *Bulletin of the History of Medicine*, 15 (1964), 180-188; see also Manfred Ullmann, *Medizin im Islam* (Leiden, 1970), pp. 137-138. Ibn al-Jazzār, *Zād al-Musāfir*, for which see Ullmann, *Medizin im Islam*, pp. 147-148, and Marcel Dugat, "Études sur le traité de médecine d'Abou Djafar Ah'mad, intitulé Zad al-mozafir, 'La provision du voyageur'," *Journal asiatique* (April-May 1853), reprint at the New York Academy of Medicine.

[28] *Breviarium Constantini dictum Viaticum*, in Rasis, *Opera parva* (Lyon, 1510), fols. 1r-102v, at the New York Academy of Medicine. Complicating matters even further, Egidius Corboliensis (d. ca. 1224) composed a *Viaticus* [*sic*] *de signis et symptomatibus aegritudinum*, ed. Valentin Rose (Leipzig, 1907).

[29] Gerardus Cremonensis, *Glose super Viaticum*, in Munich Clm 13033, 14th c., fols. 70r-145r. Gerardus Bituricensis, *Breviarium Constantini dictum Viaticum cum expositione*, in Gerardus de Solo, *Opera* (Venice, 1505), fols. 98-192, at the New York Academy of Medicine. Bernard refers to a commentary by "Guiraldus," who is erroneously identified as Gerard de Solo (d. ca. 1350) by one editor of the *Lilium* (Paris, 1542), fol. 196v, margin.

[30] It was probably also in the free translation by Constantinus that Bernard knew the treatise *De urinis* of Isaac Judeus, for which see *Il Libro delle Urine di Isacco L'Ebreo tradotto dell'Arabo in Latino da Costantino Africano*, ed. Eugenio Fontana (Paris, 1966).

[31] *Costantino l'Africano, L'Arte Universale della Medicina* (*Pantegni*), Parte I, libro 1, ed. and tr. Marco T. Malato and Umberto de Martini (Rome, 1961).

[32] See Ullmann, *Medizin im Islam*, pp. 140-146.

[33] For example, *Lilium medicine*, fols. 26r and 45v.

[34] Haly filius abbas, *Liber medicine sive Theorica Regalis dispositio* (Venice, 1492), at the New York Academy of Medicine.

[35] "Haly prima parte libri regum et Constantinus ibidem, quod idem est": *De conservacione*, 4: *Regimen sanitatis*, fol. 61r.

other works juxtaposed in the comparative Table 2 above with the exception of Henri de Mondeville's *Cyrurgia*. Perhaps the most noteworthy absentee among the authors identified by Bernard is Albucasis (Abū al-Qāsim) of Cordova, whose book on surgery (ca. 1000), translated at Toledo in the twelfth century by Gerard of Cremona, was very important not only for surgery but also for the knowledge of anatomy and the practice of clinical medicine.[36] It seems surprising, furthermore, that Bernard only once names Joannitius (that is, Ḥunain ibn Isḥāq, d. 873 or 877), the author of the *Isagoge* or introduction to Galen's *Ars parva* or *Tegni*,[37] although he cited Galen's *Tegni* at least eleven times.[38] In contrast, Joannitius was listed in two syllabi of the Montpellier faculty of medicine, of which a synopsis is given in Table 3: one was a book list for the students issued in 1309 by Pope Clement v with the advice of three former masters of medicine,[39] and the other was a curriculum for the professors, contained in the statutes drawn up in 1340 by canon lawyers at the behest of the medical *Studium*.[40] The *Isagoge* was also part of the established collection of medical textbooks which became known first as the *Ars medicine* and then as the *Articella*.[41]

Still another Arab author who was important in the textual tradition and for clinical medicine, but who was mentioned relatively seldom by Bernard, was Rasis (Abū Bakr al-Rāzī, ca. 865–ca. 925). The latter, too, was listed in the 1309 syllabus of the medical faculty, and his *Liber nonus Almansoris* was commented on at least twice in the next generation at Montpellier, as may also be seen in the final column of the synopsis.[42] One

[36] See Ullmann, *Medizin im Islam*, pp. 149-151. Although one editor of the *Lilium* (Paris, 1542), fol. 157v, added Albucasis in the margin with reference to Bernard's discussion of ear obstruction, it would require further investigation to determine whether Bernard's detailed descriptions of some surgical procedures (see pp. 150-151 below) were based on the surgery of Albucasis.

[37] See Lutfi Sa'di, "A Bio-Bibliographical Study of Hunain Ibn Ishaq al-Ibadi (Johannitius, 809-877 A.D.)," *Bulletin* of the Institute of the History of Medicine, Johns Hopkins University, 2 (1934), 409-446. Also see Ullmann, *Medizin im Islam*, pp. 115-119. An English translation of the *Isagoge* in Edward T. Withington, *Medical History from the Earliest Times* (London, 1894), pp. 387-396.

[38] See Table 4, p. 114 below.

[39] *Cartulaire de l'Université de Montpellier*, 1 (Montpellier, 1890), document 25.

[40] Ibid., document 68.

[41] The most important recent study on the genesis of this collection is by Prof. Paul O. Kristeller, "Bartholomaeus, Musandinus and the Maurus of Salerno and Other Early Commentators of the 'Articella', with a Tentative List of Texts and Manuscripts," *Italia medioevale e umanistica*, 19 (1976), 59-89.

[42] Gerardus de Solo (fl. 1335–1371), *Practica super nono Almansoris*, in Paris MS BN lat. 7062, a. 1428, fols. 32r-261r. Johannes de Tornamira, *Clarificatorium super nono Almansoris* (1365), (Lyon, 1490), at the New York Academy of Medicine.

TABLE 3. AUTHORS AND WRITINGS – MONTPELLIER FACULTY OF MEDICINE

As found in:	the *Articella*	the 1309 book list	the 1340 curriculum	Montpellier commentaries
Hippocrates				
Pronostici	×	1	2	3
Aphorismi	×	1	1	7
Regimen acutorum	×	1	2	4
Galen				
De complexionibus	×	1	1	
De malicia complexionis		1	1	2
De simplici medicina		1	1	
De morbo et accidente		2	1	1
De crisi et crit. diebus		1	1	1
De ingenio sanitatis		2	1	1
Tegni	×	1	1	3
De differentiis febrium			1	1
De regimine sanitatis			1	1
De virtutibus naturalibus	×		1	
De iuvamentis membrorum			1	
De interioribus			1	1
Ad Glauconem, de medendi methodo			1	
"*de aliis libris Galieni*"			1	
Philaretus				
De pulsibus	×			2
Theophilus				
De pulsibus et urinis	×		1	1
Joannitius				
Isagoge in artem parvam Galieni	×	1	1	4
Isaac Judeus				
(general)		1		1
De febribus	×	1	1	
De dietis universalibus	×		1	
Constantinus				
(general)	×	1		1
[Nicolaus]				
Antidotarium	×	1		2
Rasis				
(general)		1		
Ad almansorem				2
Avicenna				
(general)		1		2
Canon 1 "totus"			1	2
3			1	
4, *fen* 1-4			3	4

might be tempted to infer that Montpellier scholars became familiar with Rasis only after 1300, if it were not for the fact that his *Liber ad Almansorem* had already been translated around 1170 by Gerard of Cremona, and his *Liber continens* in 1279 by Moses Farragut.[43] Since Rasis was not included in the 1340 curriculum, his popularity seems to have fluctuated at Montpellier.[44]

Among Arabic authors, the one to whom Bernard leaned and whose popularity evidently reached a peak in the first half of the fourteenth century, was Avicenna. The degree and advance of his influence at Montpellier is reflected in the fact that in the 1309 book list his name was mentioned without any specific work and he could be replaced by the combined writings of Rasis, Isaac and Constantinus,[45] and that in the 1340 curriculum his *Canon* yielded material for one third of the prescribed courses.[46] The *Canon* is the most frequently cited source in the writings of Bernard, who in addition also quoted Avicenna's *Sextus naturalium* and *De animalibus* (under the title *De historiis animalium*) and referred to his *Cantica* through Averroës' commentary which had recently been translated at Monpellier by Ermengaud Blasius (1283).[47] Avicenna occupies a prominent place in Bernard's work, second only to that held by Galen, and his name accounts for nearly one fourth of the total number of explicit references. In many instances, however, Bernard cited Avicenna not so much for his own thought but as the interpreter of Galen, speaking "by the authority of Galen."[48] This subordination of the leading Arab sage to Greek or Galenic medicine, in addition to the fact that Bernard referred to Arabic authors only half as often as to Greek sources, suggests that his "Arabism" ought not to be overrated.

Bernard's thought, like the Montpellier tradition itself, was firmly rooted in the Hippocratic-Galenic heritage. He called Hippocrates the

[43] See Sarton, *Introduction*, 2: 833.

[44] In the Montpellier curricula of 1537 and 1550, one course on Rasis was again prescribed according to Alexandre Germain in his introduction to the *Cartulaire*, 1: 76-78.

[45] "[Baccallarii] Avicenne vel, ejus loco, Rasis ac Constantini et Ysaac libros habere, hujusmodi promocionis tempore, teneantur": *Cartulaire*, document 25.

[46] Ibid., document 68.

[47] See Lynn Thorndike, "The Date of the Translation by Ermengaud Blasius of the Work on the Quadrant by Profatius Judeus," *Isis* 26 (1937), 306-309, for the dates of Ermengaud's translations.

[48] For example, "secundum sentenciam Galeni secundum quod narrat Avicenna": *De conservacione*, 2: *De urinis*, in Vatican MS Pal. 1174, fol. 41v; and "Avicenna convenit cum Galeno": ibid., 4: *Regimen sanitatis*, fol. 64v; also, "dicit Avicenna auctoritate Galeni": ibid. and *Lilium medicine*, passim.

"first master,"[49] and cited his *Aphorismi*, *Pronostica*, *Regimen acutorum*, *Liber epidemiarum*, and referred to the Oath or *Testamentum*.[50] Of these five writings in the Hippocratic corpus, the first three were included in the *Articella* as well as in both Montpellier syllabi of 1309 and 1340. The first four, moreover, had been commented upon by Galen[51] and were well known to Bernard also in those commentaries. Bernard mentioned Galen almost as many times as all the other authorities combined, and this proportion of explicit references is significantly higher than that in the works compared with his in Table 2. In view of the generally alleged development of Galenism into dogma in the later Middle Ages, Galen's pre-eminent position in Bernard's thought deserves careful attention, especially since that preeminence was entirely in keeping with contemporary trends at Montpellier. With the exception of the *De methodo medendi ad Glauconem*,[52] all the Galenic works listed in the Montpellier book lists were also cited by Bernard, as may be seen from Table 4. Another interesting fact revealed in this table is that six of Galen's writings that are not found in either the *Articella* or the Montpellier syllabi, are nonetheless mentioned, once each, by Bernard. Several other Galenic works may lie concealed among the numerous instances where he simply referred to Galen without giving a specific citation.

The sovereign authority exercised by Galenic medicine is evident not only in the number of Bernard's references to works by Galen, but also in his reverent attitude towards "the prince of physicians."[53] Observing that Galen had considered study the shortest road to God,[54] and that his views

[49] Thus, "magister primus Ypocras docuit pronosticare": *Lilium medicine*, fol. 97r; or, "primus magistrorum omnium Ypocras docuit pronosticare iiij aphorismorum": *De conservacione*, 2: *De urinis*, fol. 33v.

[50] "Dimisit nobis Hippocrates in mandatis in testamento suo, Non doceas potionem mortiferam nec aborsum": *De tyriaca*, ed. Pagel, p. 259. It would be worth investigating, on the basis of the term *testamentum* and of the text quoted, whether Bernard knew the oath from a direct translation or through an intermediary such as Galen, Avicenna, or an early commentator.

[51] In the reading list of 1309 Galen's commentaries on the first three of these Hippocratic writings are designated as "*medicinales commentatos*": *Cartulaire*, document 25.

[52] However, it may be assumed that Bernard knew the *De methodo medendi ad Glauconem*, which deals with a subject he considered very important, namely the differences between fevers. See Carl G. Kühn, ed., *Claudii Galieni Opera Omnia* (Leipzig, 1830), vol. 11; for manuscripts and incunabula, see TK, index.

[53] "Princeps medicorum": *Lilium medicine*, fol. 64r and passim.

[54] "Secundum Galenum VII de Ingenio sanitatis nullus potest Deo melius approximare quam studendo in veritate et per veritatem": ibid., prologue, fol. 1r.

TABLE 4. GALENIC WORKS EXPLICITLY CITED BY BERNARD

	Number of references	Works by Bernard on same topic	Found in *Articella*	In Montpellier curriculum 1309	1340
De ingenio sanitatis	110	1		2	1
In aphorismos Hippocratis	59		1	(1)	
De morbo et accidente	51	(?)	1	2	1
De crisi [and] (28) *De criticis diebus* (6)	34	1	1	1	1
Regimen sanorum	31	1	1		1
De simplici medicina	25			1	1
Regimentum [or] (11) *Regimen acutorum* (8)	19	1	1	(1)	
De complexionibus	17		1	1	1
De pronosticis	17	1	1	(1)	
De interioribus	16				1
De differentiis febrium	13	(?)	1		1
Tegni	11		1	1	1
De iuvamentis membrorum	9				1
De virtutibus naturalibus	6		1		1
De malicia complexionis	3			1	1
De marasmode	1	1			
De cibis	1				
De locis affectis	1				
De motibus	1				
De tactu pulsus	1				
De voce	1				
De methodo medendi ad Glauconem			1		1
No work cited	173				
Total	600				

agreed with those of the Church,[55] Bernard practically granted Galen the halo of a saint. Calling him "God's servant," he asserted that Galen possessed a tactile sense and an intelligence that surpassed those of ordinary humans.[56] Bernard even hinted that Galen's knowledge of

[55] Thus, citing Galen's view that growth is finished at the age of twenty-one, Bernard asserted that "in hoc Galenus convenit cum ecclesia, quoniam ecclesia non precipit ieiunium nisi post annum xxi et reddit causam ne impediat augmentum": *De conservacione*, 4: *Regimen sanitatis*, fol. 69r.

[56] "Constriccionem pulsus solus Deus novit et Galenus servus eius": ibid., 3: *De pulsibus*, fol. 52r. Also, "in pulsibus autem forte sunt duo que nullus mortalis potuit cognoscere, nisi solus Galenus... [tactile knowledge is too difficult or even impossible] nisi

medicine may have made him immune to disease, when he asserted that, "*before* he knew the art of healing, Galen fell gravely ill." Bernard commented appreciatively that Galen's eventual "recovery was a great comfort to the whole world."[57]

As could be expected, Galen's views prevailed in Bernard's works over the opinions of lesser lights. For example, although Bernard admitted that "venerable Egidius did quite well in his treatment" of the subject of uroscopy, he nonetheless appealed to Galenic views when "laying to rest" several theories of the same Egidius, who is better known to us as Gilles de Corbeil (ca. 1140–1224).[58] Bernard also invoked Galen's authority when he rejected the advice of Serapion (Jachja ben Sarabi, late ninth century): he asserted that phlebotomy should be applied at the onset of sanguineous fever, "the opinion of Serapion notwithstanding, because that is what Galen wants."[59] Similarly, where the so-called "Viaticus" disagreed with Galen, Bernard refuted his views, in one case chiding him for "judging on the basis of appearance."[60]

Even when Galen appeared to contradict a major author, Bernard usually favored him. With regard to the basic disagreements between Galen and Aristotle, we have seen earlier how Bernard preferred the physicians' thesis (*viam medicorum*), championed by Galen, over that of the philosophers (*viam philosophorum*) in attributing to semen a material and substantial role in the formation of the fetus.[61] In the controversy over

---

soli Galeno qui habuit tactum delicatissimum cum ingenio claro": ibid., 2: *De urinis*, fol. 30r.

[57] "Accidit Galeno dum, antequam sciret artem sanativam, infirmatus est egritudine forte valde ... curatus est, que curacio fuit toti mundo magna consolacio": *Liber pronosticorum*, in Vatican MS Pal. 1174, fol. 87v.

[58] Thus, since Galen taught that hairs are produced by thick humours, "ideo quiescant illi versus Egidii, Forma pili similis etc." on "hairs" found in urine. Furthermore, because according to Galen "contenta crynoidalia" in the urine are not determined by the solid limbs, "ideo non possunt attestari super terciam speciem ethice sicut opinatus est Egidius, dormiant igitur illi versus, Certa fides per crynoydes...": *De conservacione*, 2: *De urinis*, fol. 49r. Printed editions give *crimnoidalia* and *crimnoides*.

[59] "... non obstante opinione Serapionis, quia Galenus vult hoc": *Lilium medicine*, fol. 4r.

[60] For example, with regard to *iectigacio* (a spasmodic way of walking), "contrarietas videtur esse inter Galenum 5 de Morbo et Viaticum, quoniam Viaticus dicit quod est motus nature et morbi ... sed Galenus quod est morbi...." But because it is similar to tremors, "quantum ad hanc apparenciam iudicans Viaticus dicebat": *Lilium medicine*, fol. 36v. Also, "ventositas intestinorum" can result from heat according to Galen, even though "oppositum videtur dicere Viaticus," ibid., fol. 68v. On the identity of "Viaticus" see pp. 108-109 above.

[61] See pp. 78-80 above.

the question whether man is ruled by the brain or by the heart, Bernard attempted to reconcile opposing viewpoints by means of subtle distinctions: whereas Aristotle located the *sensus communis* in the heart, Bernard followed Galen in assigning it to the brain, but he explained the discrepancy as due to a difference in emphasis between the two authors. He seemed uncertain, however, of the effectiveness of his conciliatory effort when he added, "and if they can be friends through this explanation, I am satisfied."[62] The same hesitation had been expressed by Rasis: "if Galen and Aristotle are of one mind on a subject, then of course, their opinion is true. When they differ, however, it is extremely difficult to decide which opinion to accept."[63] Bernard resorted to his most subtle dialectic in his attempts to solve apparent disagreements between Galen and Avicenna, for example in his chapter on the things that preclude the administration of phlebotomy.[64] In another instance he noted that, whereas Galen (*De crisi*, 2) accepted only three simple fevers, Avicenna (*Canon*, 4) added a fourth and "took strong exception (*multum scandalizatur*) to the words of Galen." After reconciling this difference by means of the scholastic distinctions between "preceding" and "joint" causes and between "becoming" and "being," Bernard mused that "it was, nevertheless, strange that Galen made such statements."[65]

In cases where Galen "seems to contradict himself," Bernard either went to great lengths to straighten out the discrepancy, or he expressed his bewilderment over it. Thus, in a complex *questio* he pointed out that Galen attributed the green color of urine to cold, in Book One, and to heat,

[62] "Sensus communis radicaliter et originaliter est in corde et ita intelligit Aristoteles. Sed manifeste est in cerebro et ita intelligit Galenus; et si ex hiis possunt esse amici placet mihi": *Lilium medicine*, fol. 29v.

[63] Cited (without reference) by Victor Robinson, "Rhazes," *Medical Life*, 28 (1921), 208. The differences between Aristotle and Galen, the subsequent corresponding positions of philosophers and physicians, and the criticism of Galen by Rasis are summarized in the remarkable study of Owsei Temkin, *Galenism; Rise and Decline of a Medical Philosophy* (Ithaca, N.Y., 1973), especially pp. 72-82 and 118-122. Also see notes 68, 74 and 76 below.

[64] *De conservacione*, 1: *De flebotomia*, in Vatican MS Pal. 1174, fol. 16v.

[65] "Contrarietas videtur inter Galenum et Avicennam ... Avicenna autem multum fuit scandalizatus de verbis Galeni ... breviter dico quod si consideremus vaporem sanguinis putrefactum, dico quod iste convertitur in coleram, quantum autem ad causam antecedentem sanguis solum putrefit, sed quantum ad [causam] coniunctam, non. Vel possumus dicere quod in fine [*sic*, fieri] sanguis facit febrem singularem, sed quantum ad factum esse, non; uno modo intellexit Galenus, alio [modo] Avicenna, et ita non est contrarietas. Sed mirabile fuit de Galeno quod talia dixit": *De conservacione*, 2: *De urinis*, fol. 43v.

in Book Two of his commentary on the Hippocratic *Prognostications*. Bernard declared that such a green color is caused by both "burning heat and death-causing cold." He marshalled the authority of Aristotle and of Avicenna as well as the observation of greening leaves and ponds in arguing for the effects of warmth, but he failed to give specific evidence for the same effects of cold. He concluded, nevertheless, that "greenness can be caused by heat or cold, and hence there is no contradiction."[66] Elsewhere, unable to solve an apparent inconsistency in Galen's discussion of eye abscesses, Bernard candidly admitted, "what ought to be said about this, I do not see, except that it is under some cloud if the letter is true as it stands." He then added, however, the reassuring comment that it seemed to be an inconsequential "argument over words."[67] It is worth noting that in his attempts to solve contradictions in Galen's works Bernard did not take into account the possibilities, such as misunderstanding or mistranslation, which had been pointed out by Moses Maimonides (1135–1204) in his *Aphorisms*.[68] Instead, when one of his own explanations might surprise the reader, he hastened to profess that "nothing of what I say here is contrary to the sayings of Galen."[69]

In spite of his avowed respect for Galen's authority and in spite of his endeavors to reconcile seeming contradictions, Bernard was reluctant to suspend or surrender his own independent and critical judgment. This is illustrated, for example, in his *questio* "whether hot water baths are suitable for athletes and wrestlers." Overruling the negative answer dictated by Galen's view that hot water might provoke fever, Bernard proceeded by describing the warm baths he would administer, but then he

[66] "Queritur de contrarietate que videtur inter Galenum et seipsum; videtur dicere quod color viridis causatur a frigore, primo pronosticorum ... oppositum videtur dicere Galenus secundo pronosticorum, quod color viridis causatur a calore, et ita videtur contrarius sibi ipsi. Dico quod color viridis causatur a calido adurente et frigido mortificante ... videns Galenus quod color viridis poterat causari et a calido et a frigido, dicebat quod causabitur a frigido et postea dicebat quod a calido, et ita nulla est contrarietas": *De conservacione*, 2: *De urinis*, fol. 46v. Another instance of apparent Galenic inconsistency is cited by Bernard in the *Lilium medicine*, fol. 88v.

[67] "Galenus videtur contrarius sibi ipsi ... quid sit dicendum ad hoc ego non video, nisi sub quodam nubilo si littera est vera sicut iacet ... quoquomodo se habeat negocium, propter ista verba non mutatur causa nec signum nec operacio et ideo videtur mihi quod disputacio sit de nominibus": *Lilium medicine*, in the printed edition (Venice, 1498), fol. 39v. This passage is lacking in London BL MS Harl. 3698, but the badly mangled text on fol. 38v there and a comparison with other early MSS leave no doubt about the authenticity of the passage.

[68] See Temkin, *Galenism*, pp. 123-124.

[69] "Nec aliquid quod hic sit dictum est contrarium dictis Galeni v de Ingenio," *Lilium medicine*, in London BL MS Harl. 3698, fol. 42v.

hesitated: "however, Galen's authority is greater than all my perspicacity; hence, if Galen's word is to be taken literally, let his statements be followed."[70] Still insisting on his own opinion, he claimed that Galen's objection actually related to hot rather than to warm water, and thus he cleverly solved the contradiction by means of semantics or "*modus loquendi*."[71] He applied similar subtlety and caution when he discredited the allegation that Galen had invented the almost miraculous antidote, theriac.[72] He was more direct, however, when he deplored Galen's "habitual neglect to assemble an entire subject matter in one place,"[73] echoing earlier criticisms of the lack of synthesis as a basic weakness in Galen's method.[74] On one occasion Bernard deferentially interjected, "*salvo honore Galeni*," when disagreeing with the latter,[75] but elsewhere he firmly asserted his own opinion as more "reasonable" than that of Galen.[76]

One prescription of Galen, namely that phlebotomy should be applied until the patient fainted, provoked Bernard's most outspoken disagreement. His objection may have been inspired by that of Averroës,[77] who was often critical of Galen,[78] but it evidently also sprang from personal conviction: it is reiterated in three of his works, each time with increased emphasis and with a more refined argumentation. In the *Lilium medicine*

[70] "Queritur utrum balneum aque ferventis competat in atletis et pugnatoribus ... et videtur quod non.... Talis aqua inducit vigorem secundum Galenum in de Morbo et ita esset timendum de febri.... Respondeo, primo dico quid ego facerem.... Attamen, quia maior est Galeni auctoritas quam tota mea perspicacitas, ideo si littera Galeni est vera sumantur dicta eius": *De conservacione*, 4: *Regimen sanitatis*, fol. 70v.

[71] "Ad aliud dico quod non est tanti caloris sicut pretendit vocabulum quod nec posset sustinere nec esset obediens, sed fuit quidam modus loquendi, scilicet quod aqua esset fervens quasi calida ultra temperamentum": ibid.

[72] "Dicitur in Antidotario quod Galenus componit eam [tyriacam], sed non est verum secundum omnes alios auctores; bene investigavit et consideravit et examinavit, sed Andromachus rex omnino ipsam perfecit antequam esset Galenus": *De tyriaca*, ed. Pagel, p. 258. The *Antidotarius* to which Bernard alludes is that by Nicolaus Salernitanus according to Pagel, ibid., n. 9.

[73] "Fuit consuetudo Galeni quod raro voluit omnino totam materiam aggregare in uno loco": *De conservacione*, 1: *De flebotomia*, fol. 12v.

[74] See, for example, Temkin, *Galenism*, p. 68.

[75] "Disenteria appellatur fluxus humoralis cum sanguine, et salvo honore Galeni bene potest apparere fluxus sanguinis cum lyenteria [i.e. fluxu ventris]": *Lilium medicine*, fol. 69r.

[76] "Secundum Galenum vii annis lactetur infans .. sed racionabile videtur iii annis [for boys] vel ii [for girls]": *De conservacione*, 4: *Regimen sanitatis*, fol. 55r.

[77] See Temkin, *Galenism*, p. 121, citing Averroës, *Colliget*, 7: 8.

[78] See Christoph Bürgel, *Averroës "contra Galenum"* in *Nachrichten der Akademie der Wissenschaften in Göttingen*, 1: *Philologisch-historische Klasse*, 9 (1967), 265-340.

he stated, "it seems to me that it is safer to phlebotomize several times until the [harmful] matter is exhausted, rather than the patient: for I prefer that the illness be somewhat prolonged rather than that a death be attributed to me."[79] In the treatise *De urinis*, to the claim that his procedure was safer, he added the argument that it was also more current in his own time.[80] In his *Regimen sanitatis*, finally, he declared that "many things, which were allowed to Galen, are not allowed to us, because Galen knew better and with more certainty the strength" of the patient.[81] With this cogent argument Bernard's Galenism appears to have come full circle. While Galen's extraordinary stature inspired Bernard with reverence, as has been indicated above, it was for him also an important reason why ordinary men could not apply literally and absolutely every Galenic dictum. This basic reservation, together with his personal elaboration of the ongoing critique of Galen, seems to warrant the conclusion that, whatever influence Galenism exerted as a medical philosophy, Bernard did not accept every word of Galen as part of an inviolable and absolute dogma.

The limits, which Bernard imposed upon his assent to Galen's views, applied a fortiori to his acceptance of other sources. Although authority and tradition did carry considerable weight, two additional criteria were needed to establish a fact beyond doubt. One such fact was the force of habit which, as Bernard stated, "is manifest not only by way of experience but also in the authorities and by reason."[82] Reason and experience,

[79] "Intencio igitur Galeni fuit IX de Ingenio quod fiat flebotomia usque ad sincopim. Et mihi videtur quod tucius est facere flebotomiam multis vicibus donec deficiat materia et non paciens. Magis enim diligo quod egritudo aliquantulum prolong[ar]etur quam si mors mihi imputaretur": *Lilium medicine*, fol. 3v. It may be noted that another Montpellier author, Gualterius Agulinus (fl. 1250), had prescribed phlebotomy "usque ad proximitatem syncopis. Cave tamen ne cadat in syncopim": *Summa medicinalis*, ed. Paul Diepgen (Leipzig, 1911), p. 101.

[80] "Galenus fecit flebotomiam usque ad syncopim [virtute stante, et nullus debet syncopen sicut dicit Galenus vereri. Medici autem nostri temporis non faciunt usque ad syncopen, sed] et hodie fiat, et cras, et ita paulatine donec materia fuerit totaliter evacuata, et iste modus est securior": *De conservacione*, 2: *De urinis*, fol. 43r. The passage in brackets is lacking in this MS, but it is found in all printed editions, including the Venice 1498 incunabulum, fol. 28r.

[81] "Ad Galenum dico quod multa erant sibi licita que non sunt nobis licita pro tanto quod cercius et melius cognoscebat virtutem et ideo precipit in XI de Ingenio facere flebotomiam usque ad sincopim; nobis autem non expediret quoniam melius est et tucius quod flebotometur bis vel ter vel amplius quam si poneremus in illo periculo": *De conservacione*, 4: *Regimen sanitatis*, fol. 68v.

[82] "Magnam igitur vim habet consuetudo, et non solum hoc est manifestum per viam experimenti ymmo et per auctoritates et per raciones, quoniam natura gaudet in consueto": *Lilium medicine*, fol. 66v.

therefore, were the touchstones by which Bernard criticized or emended several authors. In his most sweeping denunciation on the basis of reason, Bernard averred that in case of pleurisy "no strong medicine ought to be given, even though the ancients say so and some moderns boast that they follow that course. But, foregoing disputations, it yet appears not to be reasonable."[83] By "reasonable" Bernard obviously understood that which was verified either by deduction or by induction. He used deductive reasoning to refute Avicenna and "some of his followers," who had argued that "phlebotomy drains the good blood and leaves the bad blood behind." Bernard replied that "this proposition is worthless because Nature retains the good blood which she needs, and she lets the bad blood go."[84] This conclusion, though startling, was based on a solidly established philosophical principle, namely the teleological view of nature, but some of Bernard's a priori arguments were more purely a matter of speculative logic. Thus he claimed that, "even though Damascenus holds a contrary opinion, the brain and bones can be abscessed, because whatever is being fed can also swell."[85] Even more dependent upon dialectic was his refutation of the *Viaticum*'s prescription that, in the treatment of tremors, the administration of oils and laxatives ought to follow the therapy by baths. He considered this "against reason, because universal remedies [that is, baths] should precede the particular ones [that is, oils and laxatives]."[86]

Deduction and induction were welded into one blunt rebuke of Avicenna's suggestion that it was good to get drunk once a month as a complete purge. Bernard indignantly protested that "there are many other ways of purging," and that Avicenna's method was "ignominious and not even entirely safe" because it entailed the risk of apoplexy. Therefore, he concluded, "let it be completely banned from the realm of our speculation."[87] Elsewhere he injected a religious note by warning that "no

[83] "Nulla fortis medicina debet dari in pleuresi, quicquid dicant antiqui, et aliqui moderni iactant se illam viam tenere. Sed disputacionibus pretermissis non videtur racionabile": ibid., fol. 57v.

[84] "Nec videtur procedere racio quam ponit Avicenna et quidam sequaces eius, quoniam ubi est sanguis multus malus et paucus bonus, flebotomia rapit bonum et dimittit malum.... Nec valet racio, quia natura retinet bonum quod indiget, et dimittit abire malum": ibid., fol. 16v.

[85] "Cerebrum et ossa possunt apostemari, licet Damascenus teneat contrarium, quoniam quicquid nutritur potest apostemari [dilatari]": ibid., fol. 32r.

[86] "Viatici capitulo de tremore ... videtur quod balnea precedant et alia subsequantur, quod est contra racionem quia universalia debent precedere particularia": ibid., fol. 36v.

[87] "Secundum Avicennam bonum est inebriari semel in mense, non propter ebrietatem sed quia ad ebrietatem sequitur vomitus, sudor, et secessus, que corpus habent

Catholic should knowingly get drunk" and by explaining that "this method of purging the body is like a shipwreck, and it is therefore totally against God and reason."[88] Also on the subject of alcoholic beverages Bernard introduced a very pertinent conclusion of inductive reasoning, namely the argument that sometimes traditional statements did not apply because their authors had belonged to a different era, culture, or climate. Thus, Galen and Avicenna had opposed the drinking of wine by children. Bernard found their view contrary to "common usage throughout the whole world, in so far as we know, and yet children are not being harmed by such taking of wine."[89] To Galen he replied specifically that "in antiquity wine was not drunk as commonly as today: that is evident since Galen did not know the Marnau wine of Cyprus or the wine from Vernazza, for he did not mention them anywhere even though he mentions so many things at great length."[90] As for Avicenna, on the other hand, "in his time no one drank wine, in accordance with his law – they [that is, Muslims] still do not drink – and therefore he could not judge with enough certainty" on the effects of wine.[91] Incidentally, this seems to be one inconsistency which Bernard failed to notice, namely between Avicenna's Islamic prohibition and his abovementioned prescription of monthly inebriation.

There are further indications that Bernard at least sensed, but not fully grasped, the need to revise traditional doctrines because of changing historical contexts. Observing that Hippocrates, "the first master," had

mundificare.... Et cum multi alii modi sunt purgandi vel purgacionis, et hoc sit ignominiosum, et eciam non est omnino tutum, quia ante vomitum bene potest incurrere apoplexiam, ideo ab aula nostre speculacionis totaliter sit remotum": ibid., fol. 31v.

[88] "Et ideo nullus catholicus scienter inebrietur quoniam licet sequatur vomitus et sudor tamen magnum periculum est quod incurrat apostemam [*sic*, apoplexiam] aut epilenciam aut suffocacionem et ideo infiniti modi sunt meliores.... Ille autem modus purgandi corpus est sicut naufragium et ideo est totaliter contra Deum et racionem": *De conservacione*, 4: *Regimen sanitatis*, fol. 67r-v.

[89] "Queritur utrum vinum competat pueris et mittitur quod non per Galenum qui dicit in de Regimine sanitatis ... et per Avicennam.... Contrarium vult communis usus per universum orbem quantum noticia pervenit ad nos et tamen pueri ex tali accepcione vini non leduntur": ibid., fol. 58r.

[90] "In tempore antiquo non potabatur vinum ita communiter sicut nunc et patet quoniam Galenus non novit vinum de Chipro Marnau nec vinum quod vocatur de Vernachia, idque inde constat quoniam nullicubi facit mencionem de hiis licet de multis valde longam facit mencionem": ibid., fol. 58v. The same two wines were described by Bernard in his *De gradibus*, ed. Pagel, p. 260 (with annotations by the editor).

[91] "In tempore autem Avicenne nullus potabat vinum secundum legem suam nec adhuc potant et ideo non potuit determinare cum tanta certitudine": *De conservacione*, 4: *Regimen sanitatis*, fol. 58v.

taught that eunuchs do not suffer from gout (*Aphorisms*, 6.28), he admitted that this was true in so far as the disease was presumably related to sexual activity, but then he added, "today, however, the situation is different because eunuchs as well as women, [people of] both sexes and every age, follow an increasingly bad regimen, and hence they incur gout."[92] He also reiterated Galen's comment that "the sons of this era live in pleasure, self-indulgence and gluttony, and therefore they are gout-ridden."[93] Bartholomeus Anglicus, on the contrary, about 1230 ignored Galen's revision and simply asserted that eunuchs "have not podagra for they serve not Venus."[94] To be sure, the question remains whether eunuchs were sufficiently numerous in fourteenth-century Christendom to merit greater attention than they had received from Hippocrates and Galen, and that question raises some doubt about the extent of Bernard's awareness of historical perspective. In any event, he also realized the implications of climatic differences, for instance when he listed the Hippocratic prescriptions for fever and then remarked that this treatment "was administered in the time of Hippocrates; we, however, who live in another climate, do not use those medicines but rather laxatives with cassia extracts."[95] Even if they were not yet developed into a comprehensive and consistent perception of historical perspective, revisions such as those cited above illustrate a valuable application of reason to authoritative tradition.

In addition to reason, Bernard used another criterion of truth, namely experience, to supplement or to contradict his sources. In the first place he frequently supplemented traditional etiology by giving causes which "the authors do not mention, because this [cause] occurs but rarely."[96]

[92] "Magister primus Ypocras docuit pronosticare, Eunuchi non podagrizant ... et est verum quantum est ex parte coitus, sed hodie vadit aliter negocium quia eunuchi, mulieres, omnis sexus et omnis etas multiplicant malum regimen, ideo incurrunt podagram": *Lilium medicine*, fol. 97r.

[93] "Secundum Galenum [*Comm. in Aphorismis*, 6: 28] filii huius temporis vivunt in suavitate, luxuria et gulositate, et ideo sunt podagrici": ibid., fol. 97v.

[94] *De proprietatibus rerum*, 7: 59, tr. James Walsh in *Medical Life*, 40 (1933), 569.

[95] "Secundum Ypocratem competit farmacia ut elleborus niger ... vel fellichimum cum asa, et istud dabatur tempore Ypocratis. ... Nos vero, qui sumus in alio climate, non utimur illis medicinis sed ventrem lenimus cum cassiasi": *Compendium regiminis acutorum*, in Vatican MS Pal. 1083, fol. 282v.

[96] For example, colic can be caused by a stone in the intestines, "sed hoc est de raro contingentibus, ideo auctores non faciunt mencionem": *Lilium medicine*, fol. 74r. Or, "auctores communiter loquendo non ponunt nisi duas species morfee ... de quo summe miror. ... Et forte non posuerunt auctores quia non erant [cause] famose et raro accidunt": ibid., fol. 16r. Also ibid., fols. 71r, 3v, and so on.

Furthermore, Bernard recommended medications which "the authors whom we have consulted do not mention, but experience truly proves that they are most effective against poisoning."[97] To the traditional cures for a widespread intestinal affection, namely worms, he added one that "has now become a new and common practice";[98] and to the cures for dropsy he added one which he claimed to have tried, "*sum expertus*."[99] The way in which Bernard supplemented authoritative tradition with the fruits of his personal experience is most clearly exemplified in his report of an incident in his career:

> It happened to me one time that a certain very old surgeon suffered an intolerable ear ache. The universal purgings and particular treatments according to the teachings of Galen, Avicenna and all the other authors preceded, but they were utterly without effect. Then I applied camomile oil and in truth he was cured.[100]

As the anecdote above suggests, Bernard drew upon experience not only to supplement but also to question the therapeutic prescriptions of his sources; the same is true for diagnostics and prognostication. With regard to the latter, for example, he found little comfort in the traditional prognosis that apoplexy could be cured: "After Hippocrates and Galen have [thus] prognosticated, I say that I have seen many apoplectics and, in truth, I have never seen anyone escape; therefore, deliver us, Lord, from sudden and eternal death."[101] In the area of diagnostics, Bernard seemed ready to accept traditional views only when they were borne out by personal experience. For example, he initially rejected as "a fable" the principle of uroscopy that the four layers or regions in stagnant urine indicate specific groups of diseases, because that would mean that "the regions are arranged in accordance with intelligence and reason," something he deemed "utterly impossible." He confessed, however, that in the end he became "convinced because of experience (*propter*

[97] "Auctores enim quos vidimus non faciunt mencionem de istis [sc. tormentilla et herba tunici]; experimentum tamen vere certificat quod summe valent contra venenum": ibid., fol. 7v.

[98] "Ecce quod fit modo communiter de novo...": ibid., fol. 75r.

[99] "Ego autem sum expertus succum radicis yreos": ibid., fol. 81r.

[100] "Accidit mihi semel quod quidam restaurator antiquus paciebatur in auribus dolorem intolerabilem, et precesserunt universales purgaciones et operaciones particulares iuxta doctrinam Galeni et Avicenne et aliorum auctorum omnium, et nihil penitus valebant, et tunc apposui oleum de camomilla et in veritate curatus fuit": ibid., fol. 42v.

[101] "Postquam igitur Ypocras et Avicenna pronosticaverunt, dico ego quod ego vidi multos apoplecticos et in veritate nunquam vidi aliquem evadere; a morte igitur subitanea et morte eterna libera nos Domine": ibid., fol. 34v.

experienciam)," remembering "the word of Avicenna that experience (*experimentum*) prevails over reason."[102] At the same time, Bernard realized the fallibility of experience, since he admitted that "the physician is often misled," for instance into diagnosing colic as kidney trouble.[103]

When his personal experience seemed in direct conflict with the diagnostics of all the authors, including Galen and Avicenna, Bernard was in a quandary. He was especially baffled by the diagnosis of leprosy. He had once seen a patient who was apparently afflicted with the most advanced deterioration of the disease, but who showed no noticeable marks in the face. Could, then, anyone be truly leprous without any facial symptoms, even though all authors held that the signs of leprosy would show first in the face? Perplexed, Bernard admitted, "about this question, I do not know, God knows; but let me relate here without prejudice what happened to me."[104] In a preliminary attempt to reconcile his observation with tradition, he suggested that "it is quite possible that the symptoms are sometimes more detectable in parts other than the face" for several reasons, including the fact that people take better care of their face.[105] He then reported his personal observation in some detail because he felt that "nowadays lepers are diagnosed most inadequately."[106] Unfortunately, he left the adequacy of his own diagnosis in doubt, first by mentioning no other method of examination than questioning the patient, and then by relating no clinical details beyond the fact that, after almost twenty years,

[102] "Attamen ista verba aliquando fuerunt mihi vise [*sic*, visa] fabule et ideo fui frequenter contra istum versum, Et similis simili servit, et regio regioni. Quia secundum hunc modum videntur regiones ordinari secundum intellectum et racionem, quod est omnino impossibile. Attamen nunc victus sum propter *experienciam* et recordor de dicto Avicenne, Experimentum enim in hoc transit racionem ... et eciam experimento magis est credendum quam racioni cum contradicunt": *De conservacione*, 2: *De urinis*, fols. 30v-31r.

[103] "frequenter decipitur medicus": *Lilium medicine*, fol. 74r. Also, the symptoms of pestilential fever "ut plurimum deludunt medicos," ibid., fol. 6r.

[104] "Ista materia est ita sordida quod magis clarificacione indiget, et ideo si aliqualiter prolongetur nullus debet habere pro malo. In primis occurrit mihi dubium propter ea que vidi in quodam viro, utrum esset possibile quod aliquis esset perfecte leprosus, nullo signo lepre in facie apparente, et videtur quod sic.... Oppositum videtur sentire Galenus, et Avicenna et alii omnes qui ponunt signa lepre incipere a signis repertis in facie.... De ista questione ego nescio, Deus scit, sed tamen quod occurrit mihi sine preiudicio dicam hic...": ibid., fol. 14v.

[105] "Dico quod bene est possibile quod signa lepre sunt magis detecta aliquando in aliis partibus quam in facie ... quia homines sunt magis solliciti in regimine faciei": ibid., fol. 15r.

[106] "Dico vobis quod tociens repeto de corrupcione figure, quoniam pessime iudicantur ut mihi videtur hodie leprosi; qui igitur aures habet advertat si vult": ibid.

the man still lived with the same lesions of his skin and extremities yet without any sign in the face. In part because of the traditional doctrine, but perhaps even more because he "wanted to absolve" the patient from a cruel stigma, he decided to "conjecture in approximation of the truth" that the patient must have been afflicted with a combination of ailments, including arthritis and scabies, rather than with real leprosy. Although he admitted that his opinion had changed after "having worked diligently" on this case, or on the preparation of this chapter, or in the practice of medicine – it is difficult to determine exactly what he meant by "*in isto opere*" – he concluded, "I would no longer judge anyone leprous; however, God knows the truth, I do not know."[107]

If Bernard wavered when authority and experience disagreed on the subjects of prognostication and diagnostics, he seemed more confident when it came to therapeutics. Thus, presumably speaking from personal experience, he recommended phlebotomy as a treatment of hemoptysis "even though Guiraldus and some others seem to say the contrary,"[108] and he asserted more categorically that it was good for all kinds of scabies "even though the authors say the contrary."[109] He also objected to Avicenna's preference for medicine above surgery in case of dropsy: "I have never seen anyone completely cured by this method without eventually dying. I believe that it would be better to cauterize five or six times below the navel."[110] He even seemed to challenge an entire tradition when he warned against the use of narcotics for ringing in the ears, "whatever the authors may say" and "whatever may have been said or will still be said in the future."[111] It is of interest to indicate here that

[107] "Ad exemplum dico quantum occurrit mihi quod non erat lepra sed erat quedam species arthetice et quedam leprositas et scabies et ulcus unguium.... Et ideo dico vobis quod illum volui absolvere et frequenter interrogavi si aliquod signum apparebat in facie, et steterat sic fere xx annis, et adhuc vivit cum illa feditate extremorum sed nihil apparet in facie. Et ideo coniector coniectura proxima veritati quod non erat lepra nec videtur possibile quod tanto tempore stetisset quin facies fuisset corrupta. Et ideo licet fuissem alterius opinionis, modo cum in isto opere diligenter laboraverim sum alterius, nec modo aliquem leprosum iudicarem; Deus tamen scit veritatem, ego nescio": ibid.

[108] "... licet Guiraldus et quidam alii videantur dicere oppositum": ibid., fol. 54r. For the identity of Guiraldus, see n. 29 above.

[109] "Flebotomia competit ... licet auctores dicant oppositum": *Lilium medicine*, fol. 16v.

[110] "Nunquam vidi aliquem perfecte curatum per hunc modum quin finaliter moreretur. Credo quod melius esset quod fieret v vel vi cauteria subtus umbilicum": ibid., fol. 81r.

[111] "Quicquid dicant auctores, melius est dimittere narcotica ... quicquid dictum sit et ex hoc eciam dicetur": ibid., fol. 41v.

Bernard frequently cautioned that narcotics should be "avoided as much as possible."[112] While for headaches he listed several painkillers (*narcotizancia*) derived from opium or papaver, henbane, and mandrake, he warned that they "should not be applied except in great necessity, as it were in desperation."[113] Against pain in general, he taught that narcotics could be used "if we have no other way than to induce stupor," but with due caution for the danger of overdose.[114] Since he knew the lethal qualities of opiates,[115] his warnings may be read as specific applications of the pledge in the Hippocratic oath, "I will give no deadly medicine to any one if asked, nor suggest any such counsel."[116]

Bernard emphasized the paramount importance of experience not only in selecting one form of therapy, such as phlebotomy, and rejecting another, such as the use of narcotics, but also in determining the very effectiveness of a particular drug. This emphasis was in keeping with the Hippocratic view that "it is a valuable piece of knowledge to learn what things have been tried and have proved ineffectual, and wherefore they did not succeed."[117] In one instance he noted with some amazement that, even though most authors recommended balsam oil for its warming effects, Galen allegedly had "tried it (*expertus est*)" but had "not found it helpful": hence he, Bernard himself, advised that "it be left to verification by experience (*veritati experimenti*)."[118] In general he warned that all prescriptions, "unless tested by experience, are speeches of rhetoricians who offer fancy propositions that have nothing to do with science."[119] More specifically, he taught that both the nature of a drug and the

[112] For example, "narcotica debent fugari quantum est possibile" in case of toothache: ibid., fol. 47r.

[113] "Cavendum tamen quod ista narcotizancia non exhibeantur nisi in magna necessitate et quasi in desperacione": ibid., fol. 24v.

[114] "Si non est nobis alia via, quod stuporem inducamus, sed hoc fiat cum cautela quod in stupefaciendo non transeamus": *De ingenio sanitatis*, in Vatican MS Pal. 1083, fol. 283r.

[115] In case of diarrhea, "quidam supposuerunt sibi aliquando opium et mortui sunt": *Lilium medicine*, fol. 73r.

[116] Tr. Francis Adams, in *The Theory and Practice of Medicine by Hippocrates*, ed. Emerson Kelly (New York, 1964), p. x. This pledge is, in fact, cited by Bernard: "caveat tamen sibi quicumque sub pena sempiterna quod non doceat pocionem mortiferam nec aborsum facientem": *Lilium medicine*, fol. 95v.

[117] *On the Articulations*, 47, tr. Adams, in *Theory and Practice*, ed. Kelly, p. 239.

[118] "Dicit Damascenus auctoritate Galeni quod Galenus expertus est oleum de balsamo ubi calefaccio est necessaria, et non invenit iuvamentum, et est mirabile cum sit calidum valde et preciosum, et cum multum ab auctoribus commendetur; veritati igitur experimenti relinquatur": *Lilium medicine*, fol. 25r.

[119] "Nisi totum [medicamen] postea experimento probetur, oraciones sunt rhetoricorum, qui proposiciones habent decoratissimas scienciam non tangentes": ibid., fol. 18r.

intensity of its effect were doubtful without experiential verification: "any consideration, no matter how reasonable, has no value unless it is proven by experience (*nisi comprobetur experimento*)," which indicates or "demonstrates (*clarificat*) not only the quantity but also the quality."[120] Theriac, for example, "before it is applied to the human body, should be tested as to whether it has the proper form."[121] One of Bernard's "testing methods (*modus probandi*)" is of particular interest. Recommending the pheasant as the best test animal "on account of its greater dryness" of complexion, he prescribed the following procedure:

> Take two pheasants, cut off their crests, apply to the wounds (or administer *per os*) a toxic substance, and wait until they begin to stagger. Then put theriac on the crest wound and in the drink of one of them: if this one lives and the other dies, the theriac is good.[122]

Bernard was evidently aware of the special care required in order to know a drug's effect "*per viam experimenti*." Citing Galen's work *De simplici medicina*,[123] he stated five conditions to be observed: the substance should be free of any "accidentally acquired" or secondary qualities, "otherwise the test (*experimentum*) would be fallacious"; it should be homogeneous; it is to be applied to a "temperate" human body; or its strength should be adjusted in proportion opposite to the body's distemper "so that no error may occur"; and its secondary (*per accidens*) effect should not be confused with its primary (*per se*) effect.[124] Once more

[120] "Recurrendum igitur ad experimentum quod omnia rectificat ... nec valet aliqua consideracio quantumcumque racionabilis fuerit nisi experimento comprobetur; et sicut dico quod experimentum clarificat quantitatem, ita et qualitatem": ibid., fol. 66r.

[121] "Antequam tyriaca exhibeatur corpori humano, probanda est, si habet formam propriam": *De tyriaca*, ed. Pagel, p. 260. It may be noted that the chapter's title, "De modis probandis ipsam tyriacam," was translated by Pagel as "Experimentelles" (ibid.).

[122] "Accipiatur gallus silvester qui fasianus appellatur, quia et ille est melior propter majorem siccitatem; tunc accipiantur duo et abscindantur cristae et tunc in locis abscisis apponatur toxicum vel detur per os et tunc dimittantur donec incipiant titubare; tunc alteri ponatur tyriaca supra cristam et cum hoc detur sibi in potu, tunc si ille vivat et alter moriatur bona est tyriaca": ibid.

[123] See *Galieni ... Opera* (Venice, 1490), fol. 245r. A comparison between the text in this translation (rather than that in the Kühn edition, 11: 456, which was not available in the early fourteenth century) and Bernard's version indicates that Bernard modified some of Galen's guidelines.

[124] "Secundum Galenum in de Simplici medicina ad hoc quod aliquid cognoscamus per viam experimenti oportet 5 condiciones observare: 1ª est haec, quod res sit denudata ab omni qualitate accidentaliter acquisita ... aliter experimentum erit fallax. – 2ª condicio: quod medicina sit homogenea.... – 3ª: quod applicetur corpori humano temperato... – 4ª condicio: quod si medicina applicata sit corpori distemperato, quod medicina elevetur in gradu opposito ... ut ergo non fiat error. ... – 5ª condicio est: si medicina habeat duos

supplementing his authoritative source, Bernard then added a sixth condition of his own:

> If we wish to try (*experiri*) a certain medicine on the human body, we should try it first on birds, next on mammals, then in hospitals, then on lesser brethren and then on others in [ascending] order, because if it should perhaps be of the class of poisons it would kill.[125]

Turning next to the procedure for measuring the "degree" or intensity of a drug's efficacy, Bernard laid down five ground rules with the "frequently repeated application" of the drug as a dominant theme.[126] This theme, together with the key ideas in the conditions for testing cited above, may evoke at least one characteristic of the truly "scientific" experiment, namely controlled verification. In other respects, however, Bernard's teaching reminds us of certain important differences between what he understood by *experimentum* and what we now define as an "experiment." A first contrast lies in his concept of verification itself, of its object, and of its relationship with reason. A second and complementary difference will be found in his use of the term *experimentum* for a wide range of things, from the carefully observed "proof" to the casual "test" and from the fortuitously discovered "recipe" to all sense "experience."

These differences in interpretation may be delineated further at the hand of a recent observation by a scholar who has devoted careful study to medieval medicine and science. Professor Michael McVaugh has observed that in thirteenth- and fourteenth-century collections of *Experimenta* the word *experimentum* referred to "an event which was purely contingent, which could be known only by directly experiencing it and which could not be rationally foreseen."[127] Professor McVaugh was

---

effectus, quorum unus est per se et alter per accidens, quod effectus per accidens attribuatur effectui per se et non e contrario...": *De gradibus*, ed. Pagel, p. 143.

[125] "Et possumus addere 6^{am} condicionem gratia doctrinae: quod si aliquam medicinam volumus experiri in corpore humano, quod prius experiamur in avibus et postea in brutis animalibus et postea in hospitalibus et postea in fratribus minoribus et postea in aliis per ordinem, quia si forte esset de genere venenorum interficeret": ibid.

[126] "Ponamus talem regulam, omnis enim medicina quae multis vicibus applicata corpori humano non alteret ... erit temperata. —2^{a} regula est talis: omnis medicina applicata ... multis vicibus quae in quantum invenitur transmutare corpus quasi aliqua transmutatione occulta quae quidem per multam et continuam applicationem manifestatur, talis erit juxta naturam suam calida aut frigida aut sicca aut humida in primo gradu...": ibid.

[127] Michael R. McVaugh, "Two Montpellier Recipe Collections," *Manuscripta*, 20 (1976), 178 and n. 5.

in this case discussing medical recipes, but his assessment sheds at least indirect light on the other things denoted by the term *experimentum*, including the "proofs" for which Bernard's procedures were outlined above. Those procedures, too, were designed to determine a contingency, such as the fact that a particular theriac preparation had its "proper form." They were also predicated on a datum beyond rational definition, in this instance the "virtue" or "beneficial character (*aspectus benevolus*)" by which theriac presumably reinforced the body's natural heat.[128] Similarly Bernard's experiential method or *via experimenti* consisted of conditions and rules for the measuring of a drug's intensity but it did not demonstrate the relation between quantity and quality as an element of the rational order of nature. The rational method or *via rationis*, on the other hand, consisted primarily in the application of mathematics to the experientially acquired knowledge, but it did not reflect upon the connection between this knowledge and nature's laws. The inability to integrate more fully both methods was an intellectual disappointment ("*amaritudo intellectus*") for Bernard.[129] This inability, however, should not obscure either the contributions made in this area by Bernard (and by his contemporaries)[130] or his attempts to combine reason with experience and to seek in nature the ultimate key to intelligibility.[131]

Bernard's concern with the combination of reason and experience is further documented in those instances where he applied the term *experimentum* to things whose affinity with the modern experiment is more tenuous than that of the controlled proof, such as the casual test, the practical recipe, and general sense experience. When he used the term in reference to a casual test, for example, he apparently required certain rational standards of validity since he did not accept all the traditional tests that were supposed to determine the fertility of women and men. He recommended one test by aromatic suffumigation, taught by Hippocrates, as valid ("*valet*") for a virgin; and a more drastic one by garlic

[128] Bernard de Gordon, *De tyriaca*, ed. Pagel, pp. 223, 236, 257, and 259; also see n. 121 above.

[129] *De gradibus*, ed. Pagel, pp. 131-132; see also my earlier discussion of this treatise, pp. 49-51 above.

[130] Those contributions and their limitations are carefully weighed by Michael McVaugh in the scholarly study that accompanies his edition of *Arnaldi de Villanova Opera Medica Omnia*, 2: *Aphorismi de Gradibus* (Granada/Barcelona, 1975), especially pp. 25-48, 75-87, and 123-127.

[131] For the place of nature in Bernard's etiology, therapeutics and ethics, see chapter 5 below. Some broader implications are examined in my paper, "Nature and the Art of Medicine in the Later Middle Ages," *Mediaevalia*, 2 (1976), 23-47.

suppository, prescribed by "some authors," as appropriate for the non-virgin or "*mulier corrupta*."[132] He also cited a test for male fertility based on the belief that unhealthy sperm floats on water,[133] an opinion of Aristotle which had been contradicted by Avicenna (*Canon* 3, *fen* 21, tract. 1, cap. 9) but adopted by Constantinus Africanus (*Pantechne*, Theory 9, cap. 13).[134] Bernard added the observation, however, that "many other tests (*experimenta*) are given by the authors, but I do not attach any faith to them, because they seem to be quack tests (*empirica*)."[135]

A distinction like that between *experimenta* and *empirica* was made by Bernard not only in diagnostic tests but also in therapeutic recipes, and although it was neither fully explained nor consistently maintained, it allows us to gain a clearer idea of his criteria. His most unequivocal statement occurs in his outline of the practitioner's *modus operandi*, which will be examined in more detail in the following chapter. In that statement he mentioned "all quackeries (*omnia empirica*)" in one breath with such crafts as soothsaying and sorcery, and he declared that they were "condemned" by virtue of the physician's obligation to "operate in accordance with the art, that is by the rules (*per canones*) of Galen, Hippocrates, and the other sages."[136] Therapeutic methods, therefore, were to be judged primarily in relation to the "art" of medicine or, more specifically, to the Hippocratic-Galenic "canons" – a term which implies the importance of both established tradition and rational discipline.

Some further aspects of Bernard's distinction in therapeutic methods are implied in his treatment of tertian fever. After summarizing the appropriate regimen and other conventional prescriptions he continued, "because the young greatly enjoy experimental recipes (*experimenta*), let

[132] "Ypocras enim docuit experimentum utrum mulier esset parata conceptui...; hoc igitur experimentum valet in virginibus. Si autem mulier fuerit corrupta dicunt aliqui auctores quod supponatur sibi dolsam [?] allii et si ab intus senciat odorem allii ascendere usque ad os et nares parata est ad conceptum": *Lilium medicine*, fol. 94v.

[133] "Tercio potest experiri sic, quod semina proiiciantur in scutella plena aqua frigida, et si supernatet ille est sterilis, si vadat ad fundum fecundus indicatur": ibid.

[134] See Paul Diepgen, *Frau und Frauenheilkunde in der Kultur des Mittelalters* (Stuttgart, 1963), p. 172.

[135] "Multa alia experimenta ponuntur ab auctoribus, sed non adhibeo fidem quia empirica videntur": *Lilium medicine*, fol. 94v.

[136] "[Medicus] operetur secundum artem scilicet per canones Galeni et Ypocratis et aliorum sapiencium, et per hoc condempnatur omnis ars augurandi, vaticinandi et divinandi, sicut est ars geometrialis et suspendendo herbas ad collum, et omnia empirica et sorticina et fassina, et alia plurima que non est bonum revelare propter abutentes": *De ingenio sanitatis*, fol. 283v. See pp. 157-159 below.

us give some here." Among these recipes he listed "nettle seed with salt which, applied to the brachial artery, cures all tertian fever," and devil's-bit or blazing star (a plant of the genus *Scabiosa*) which "cures all tertian fever by its specific power (*a virtute specifica*)." He immediately added the warning that "safer than all these, however, is to proceed by the method described above."[137] Hence it is evident that he deemed the experimental recipes less reliable than the conventional treatment and that he included them only as supplements, in this case for the instruction of the young or, more frequently as we shall see later, for diseases that frustrated rational prescriptions. It is further worth noting that, in attributing a "specific power" to the last mentioned recipe, Bernard was using one of the technical terms which referred to a very special quality found in some substances. That quality might be medicinal or of another type, such as magnetism; derived from the heavens, it could not be rationally explained and thus it was more commonly called "occult power."[138]

A third text that sheds light on the relationship between experimental and rational prescriptions is Bernard's discussion of the cures for quinsy, a disease that is "most dangerous" and "very acute" and therefore requires the physician's special diligence.[139] The impressive array of conventional methods, ranging from diet and phlebotomy to compresses and gargles, is succeeded by a brief addendum:

> The following, however, work by their entire species (*a tota specie*) as our masters say: the droppings of a swallow and the excrement of a child, of a dog that is eating bones, and of a wolf. ... They also say that if a viper be strangled with a thread and be wrapped around the neck, it is very helpful. I have not tried these *empirica*, whereas I have applied the other mentioned [remedies] frequently.[140]

[137] "Et quia iuvenes multum gaudent in experimentis, ideo aliqua ponamus. Semen urtice tritum cum sale et positum supra arteriam in brachio, curat omnem tertianam. ... Morsus autem diaboli curat omnem tertianam a virtute specifica. In omnibus tamen istis securius esset incedere per viam supradictam": *De conservacione*, 2: *De urinis*, fol. 42v.

[138] Thus, Bernard stated that a poison could be lethal when "ex aliquo aspectu orbis acquirit virtutem occultam per quam operatur": *Lilium medicine*, fol. 7v. For other terms, such as *proprietas* and *forma specifica*, used by Arnold of Villanova, see McVaugh, *Arnaldi de Villanova ... Aphorismi de Gradibus*, p. 118.

[139] "Ista egritudo est multum periculosum. ... In curacione istorum apostematum non debet medicus esse negligens nec sompniculosus quia passio est acutissima": *Lilium medicine*, fol. 48v.

[140] "Illa autem que valent a tota specie sicut dicunt magistri nostri sunt stercus yrundini et stercus pueri et stercus canis comedentis ossa et stercus lupi. ... Dicunt eciam quod si vipera stranguletur cum filo et circumdetur collum, quod multum valet. Ista empirica non sum expertus, sed alia que dicta sunt feci frequenter": ibid.

Here again, Bernard apparently made a distinction between the *empirica* and the regular remedies. For the *empirica*, which were said to work entirely by virtue of their own "specific" or occult power, he deferred to the authority of unidentified "masters." This presentation seemed to imply not only the reverence but also the aloofness of the colloquial "if they say so"; in view of this implication it is of interest to note that Bernard, unlike other authors in similar cases, did not expressly refer to widely available collections of recipes such as the *Liber experimentorum* attributed to Galen and to Rasis.[141] The regular remedies, on the other hand, were presented as used by Bernard personally. They were also correlated with the particular cause of the disease, as their ingredients were adjusted according to whether the quinsy resulted from a "cold" or a "hot" cause.[142] These remedies, therefore, were based on a combination of experiential and rational knowledge: even if their powers might be known only by experience, their application depended on reason's deduction from such paradigms as the primary qualities (hot/cold, dry/moist) and cures by contraries.

The close correlation between the section on therapy and the section on etiology in this chapter on quinsy, as in virtually every chapter of the *Lilium medicine*, illustrates the paramount importance of the knowledge of causes in the rational practice of medicine. That importance was strongly underscored in Bernard's admonition that the physician should for every disease "first of all pay attention to the cause, because if you are ignorant of the cause how can you effect a cure?"[143] The knowledge of causes as prerequisite to "experimental" treatment is probably also what he had in mind when he stated that "if certain medications do not work, it is beneficial to apply many and diverse [remedies], provided they be regulated by reason."[144] In the same vein he warned that "the light of reason" should precede the "*experimentum*" and guide the "*experimentator*."[145] The "experimenter," by whom Bernard seemed to mean here the

[141] See Thorndike, *Magic*, 2: 751-758 and his entire chapter lxiv.

[142] Thus, "in primis fiat gargarisma si causa est calida ... et si causa esset frigida fiat decoccio de vino...": *Lilium medicine*, fol. 48v.

[143] "In primis hic et in quolibet morbo intendimus ad causam, quoniam si causam ignorares quomodo curares": ibid., fol. 4v.

[144] "Cum medicamina non valent, prodest multa et diversa apponere, dum tamen sint regulata racione": *Lilium medicine*, fol. 42v.

[145] "Experimentum ... est multum utile postquam lux racionis precessit et quia omnia medicinalia que dicta sunt, sunt abstersiva et mundificativa... [and some are weak, some strong], ideo experimentator debet accipere aliqua ex fortissimis in parva quantitate et de debilibus in magna quantitate": ibid., fol. 11r.

practitioner in general but elsewhere more specifically the empiric,[146] may remind us of the "worker of experiments (*opifex experimentorum*)" whom Galen contrasted with the "author of reasoning (*auctor ratiocinationis*)."[147] In a sense, however, Bernard was endeavoring to overcome that contrast by combining rational guidance with experiential wisdom in his therapeutics, even though he was unable to avoid substantial inconsistency or to solve every paradox.[148]

A similar endeavor flawed by inconsistency characterizes Bernard's attitude towards *experimentum* in the sense of reported experience or recorded observation,[149] especially of strange phenomena. A first paradox lies in the fact that, even though he was careful to introduce hearsay evidence with explicit qualifications such as "it is said" or a past tense, he would nevertheless include it in the "experience" which, according to Avicenna's maxim, "prevails over reason."[150] Occasionally he would even use it as a major premise in arguing for an idea that seemed rationally untenable. One example was his argument for the theory that the patient's *spiritus* remains in the urine and thus allows for diagnosis up to one hour after emission:

> The fact that, in some way, the *spiritus* has its own regimen outside the body is evident from this, that a severed foot moves for a certain time. And it has been heard that a word, once started, was continued after beheading; the spirit of an evil man injures children and birds; a menstruating woman tarnishes a mirror; and a wolf makes a man hoarse. Hence it is not incongruous if the warmth and spirit of urine ... retain a certain regimen. Experience, therefore, prevails over reason if reason should fail, to the

[146] Thus, "dicunt experimentatores quod in ista passione [sc. tortura intestinorum] multum valet ericius assus...": ibid., fol. 68v. For an anonymous work with the title *Experimentator*, cited in the 13th century by Petrus Hispanus and by Thomas of Cantimpré, see Thorndike, *Magic*, 2: 767.

[147] "Modus primus harum duarum doctrinarum est proprius opificibus experimentorum, iste autem modus secundus auctoribus ratiocinationis magis est conveniens": in *Galieni ... Opera* (Venice, 1490), fol. 266r.

[148] Insufficient allowance for endeavors such as Bernard's seems to be made in the revisionist (i.e. anti-scholastic) but stimulating article by John M. Riddle, "Theory and Practice in Medieval Medicine," *Viator*, 5 (1974), 157-184 (especially pp. 170-184 for the later Middle Ages).

[149] For Bernard's attitude towards personal experience (*experiencia*) and popular opinion or lore (*opinio vulgaris*) see chapter 5 below.

[150] "Recordor de dicto Avicenne, Experimentum enim in hoc transit racionem ... et experimento magis credendum est quam racioni, cum contradicunt": *De conservacione*, 2: *De urinis*, fols. 30v-31r.

effect that the contents and appearances in diverse regions of urine give respective indications.[151]

On the basis of the same hearsay evidence Bernard had earlier argued that "it is not so strange that the basilisk, a most poisonous animal, corrupts the air and kills over a long distance."[152]

A second paradox becomes apparent in the arguments just cited: on the one hand, Bernard attached faith to mysterious phenomena because they were recorded in the authoritative tradition and thus were part of supra-rational "experience"; on the other hand, he was anxious to make them less "strange," "incongruous," or contrary to the rational order of nature. Thus, he accepted on the authority of Avicenna the belief that concentrated moonbeams could cause hemoptysis[153] and he added, "the experienced (*experti*) say that, if a moonbeam enters through a window and touches a wounded horse, it causes its death." He offered, however, a rational explanation based on the natural theory of the "excessive humidity" of moonlight.[154] (It may strike us, who have seen men walk on the arid lunar desert, as odd that the moon's "complexion" was long thought to be cold and humid.) Bernard also took for granted another traditional belief, namely that comets and falling stars announced future

[151] "Quod autem spiritus habeat aliquo modo regimen extra corpus, hoc patet, quia pes abscisus movetur aliquo tempore. Et auditum est quod verbum conceptum post abscissionem capitis proferebatur; spiritus alicuius viri malivoli ledit pueros et aves; mulier menstruata inficit speculum; et lupus facit hominem raucum. Non est igitur inconveniens si calor et spiritus qui sunt cum urina, si cum ambo sint extra et similiter et sint bene conservata, si ad tempus habent aliquod regimen. Experimentum ergo, si racio deficeret, vincit racionem, quod contenta et apparencia in diversis regionibus urine supra diversa attestantur": ibid., fol. 31r-v.

[152] "Mirabile videtur de basilisco, quod solo visu interficiat. ... Dico quod multa possunt ab animalibus emitti mala et corrupta, et ideo mulier menstruata inficit speculum et lupus facit hominem raucum. Aliqui sunt homines qui interficiunt fixe intuendo aves, pueros parvos et equos, tamquam si essent fascinatores. Et ideo non est mirabile si basiliscus, cum sit animal venenosissimum, corrumpat aerem et interficiat ad longam distanciam": *Lilium medicine*, fol. 8r. On the tradition of the basilisk and its deadly glance, from Pliny to Oresme, see Thorndike, *Magic*, 1: 75; 2: 202, 433, 562, 901, 905, and 3: 433, 435. A list of fabulous poisonous beasts is found in Bernard's *De tyriaca*, ed. Pagel, p. 235.

[153] On the probable source and the development of this belief see Brian Lawn, *The Salernitan Questions* (Oxford, 1963), p. 204. For its occurrence in Alexander Neckam's *De naturis rerum* (ca. 1200) and in Thomas of Cantimpré's *De natura rerum* (ca. 1260), see Thorndike, *Magic*, 2: 202.

[154] "Dormire in radiis lune et potissime quando intrat per foramen angustum, pessimum, ymmo facit screare sanguinem sicut dicit Avicenna, propter nimiam humiditatem. Et dicunt experti quod si radius lune intraverit per fenestram et tetigerit roncinum vulneratum, est causa mortis eiusdem": *Lilium medicine*, fol. 53r.

pestilences and even wars.[155] But again he made the suggestion of a natural process, the corruption of water and air and, "possibly, a multiplication of bile, which leads to wrath and violence."[156]

Nature, rationality, and the other mainsprings of Bernard's teaching may be delineated more sharply by a brief comparison with the tenets of Nicholas de Polonia, a Montpellier alumnus who in the last quarter of the thirteenth century became an internationally celebrated empiric (as he assures us) and an outspoken opponent of academic medicine. The "art" which for Bernard was anchored in the Hippocratic-Galenic "canon" and in the authoritative though not infallible learned tradition, was rejected by Nicholas in a versified pamphlet whose principal target he revealed in the title, *Antipocras*.[157] Sneering at his second target Nicholas bragged, "I show that a cure can well be effected without knowledge of the cause, Galen."[158] In addition to this theme, which was reiterated throughout the pamphlet, the exclusive authority of experience ("*quod ex usu expertum est*")[159] and the appeal to "faith" in remedies[160] present a sharp contrast to Bernard's view of experience guided by reason and related to causality. Finally, whereas Bernard preferred rational cures and sought natural explanations even in mysterious phenomena, Nicholas attributed the

155 "Febres pestilenciales sunt iste que adveniunt in temporibus corruptis et sterilibus cum corrumpuntur segetes et aqua et aer... [the above *incipit* is listed in TK 550]... Signa autem pestilencie future cum apparet stella comata, caudela [*sic*, possibly candela] rotunda...": *Lilium medicine*, fol. 6r.

156 "Porcio sicca [ventus] que ultra medium [aeris] elevatur est materia qua inflammantur stelle comate, caudelle (~~candel~~e Erfurt Amplon. F.267 and Vat. Pal. 1174) rotunde cadentes et assub. (asup. Vat. Pal. 1174; azub Wiesbaden 56; ad sup. Erfurt Amplon. F.267) et aliarum impressionum multarum inflammatarum apparencium in tempore calido et sicco cum fontes et pluvia (flumina Erfurt Amplon. F.267; paludes Vat. Pal. 1174) desiccantur; et tunc gentes pronosticant guerras et mortem regum, quod posset esse propter multiplicacionem colere anno illo, que movet ad iram et impetum": *Liber pronosticorum*, in Erfurt MS Amplon. F.237, fol. 79r. Some MSS (Erfurt Amplon. Q.227, Paris BN 16189) lack this passage.

157 "Incipit Antipocras, quem composuit et similiter noncupavit frater Nicholaus fratrum predicatorum. Alio autem nomine appellatur liber empiricorum. Quoniam hactenus alii ... videntes me preter artem et morem communem medicorum in morborum curationibus pillulis ac medicinis inusitatis uti...": ed. Karl Sudhoff, *Archiv für Geschichte der Medizin*, 9 (1916), 40.

158 "Hoc in sermone seu [*sic* Sudhoff; read sine] cause cognitione/Monstro, posse bene curam fieri, Galiene": ibid., p. 41, vv. 12-13. See also ibid., p. 46, vv. 202 and 227; p. 48, v. 287; and p. 51, v. 390.

159 "In eadem pagina ... non sit aliud quicquam insertum, nisi quod ex usu expertum est, frequentia et soliditate experientie sufficienter firmatum": ibid., p. 40.

160 Thus, a magic ring says, "Crede, probata loquor, corroboro cum cerebro cor...": ibid., p. 49, v. 312.

greatest efficacy to the most irrational and revolting recipes[161] and invoked divine intervention as the supernatural source of occult power.[162] This mysticism should not be overlooked if one is tempted to consider Nicholas as a harbinger of experimental science.[163] If, on the other hand, one may be inclined – not altogether without cause – to see in Bernard a representative of the traditionalism and rationalism of scholastic medicine, some corrective evidence may be found not only in his critical spirit and emphasis on experience, illustrated in this chapter, but also in his assiduous concern with medical practice, which we will examine in the following chapter.

[161] Especially ibid., p. 43, vv. 105-139, and p. 47, vv. 240-259.

[162] Thus, "sic dat bonitas deitatis" (ibid., p. 41, v. 38); also, "Non opinor pro me, sed agit deus hoc opus ex me" (ibid., p. 49, v. 323).

[163] Such a consideration seems implicit in Stanislav Szpilczynski, "Considérations sur les conceptions pseudo-scientifiques des pratiques de Nicolas de Pologne," *xvi*[e] *Congrès international d'histoire de la médecine, Résumé des communications* (Montpellier, 1958), pp. 30-31; also in Riddle, "Theory and Practice," p. 181.

5

Standards of Practice; Professional and Personal Ethics

In his treatise on the principles of medical procedure, *De decem ingeniis curandorum morborum* (1299), Bernard devoted special attention to the qualifications which he deemed essential in the practitioner. He enumerated these qualifications in one pithy sentence:

> in addition to the requirements which Haly sets down in the *Tegni*, namely good perception in the five senses, quick skill, a good memory, good judgment and sound intellect, he ought to have three other characteristics, that is to say he should be knowing and diligent, and he should practice in accordance with the art.[1]

This epitome provides us with a logical framework for the examination of Bernard's practical concerns as evinced not only in his ideas on professional requirements and on ethical principles but also in his references to personal experience and to the physician's modus operandi. The three characteristics added by Bernard may serve as our main guideposts because each of them relates to a different facet of medical practice as, respectively, the application of a science, the service of health and life, and the profession of an art.

The requirement which Bernard considered "most necessary" for the adequate practice of medicine was that the physician must be "knowing," by which he meant "able to interpret the sayings of Galen and Hippocrates, to distinguish universals from particulars, and to apply

[1] "Et cum hoc advertendum est ad medicum et preter condiciones quas ponit Haly in tegni scilicet quod sit bone visionis in quinque sensibus, et velocis sollercie, bone memoracionis, boni iudicii et boni intellectus, oportet quod habeat in se tria, quod scilicet sit sapiens, diligens, et operetur secundum artem": *De decem ingeniis curandorum morborum*, in Vatican MS Pal. 1083, fol. 283v. Instead of *sapiens* we read *sciens* in Cues MS 308, 14th c., fol. 106va (variant readings in this MS, when relevant, will be indicated in parentheses (C) in the following notes).

universals to particulars."[2] This definition reveals the close connection between theory and practice. The "interpretation" of the authoritative tradition involved not only its speculative exegesis, which was one of the teacher's chief responsibilities,[3] but also its concrete application. Furthermore, the physician's "knowledge" as defined here entailed not only his mastery of the "universals" of both *medicina theorica* and *medicina practica* but also his ability to discriminate between universal doctrines and particular observations and to relate general principles to individual cases. Thus Bernard regularly concluded his therapeutic guidelines with the caution that "the wise physician" would often have to adjust the prescribed *modus operandi* to different circumstances.[4] Conversely, the understanding of the major branches of *medicina theorica*, including pathology, anatomy and physiology, was presented as the foundation of therapy in the treatise *De ingeniis curandorum morborum*;[5] and its dominant theme was that "in everything the physician should observe what is needed in accordance with the nature of the disease and of the other particulars."[6] In the same way, Bernard's emphasis on the knowledge of causes was inspired by his desire to provide a rational basis for medical practice, as we have seen in the preceding chapter.[7]

It is of interest to note that Bernard seemed to focus on the need to know the nature and causes of disease and cure rather than on the practitioner's familiarity with anatomy. This focus was undoubtedly

[2] "Prima istarum condicionum est valde necessaria, ex hoc enim erit potens interpretari sermones Galeni et Ypocratis, et universalia a particularibus discernere, et universalia particularibus applicare": *De decem ingeniis*, in Vatican MS Pal. 1083, fol. 283v. We read *summe necessaria* in the incunabulum (Venice, 1498, at the New York Academy of Medicine), fol. 1ra (variant readings in this incunabulum will be indicated in parentheses (P) in the following notes).

[3] See chapter 4, p. 102 above.

[4] For example, "sapiens igitur medicus operabitur cum hiis et diversimode iuxta convenienciam particularium que sibi occurrunt cum omnia secundum Galenum certis litteris denotari non possunt," *Lilium medicine*, in London BL MS Harl. 3698, fol. 41v. And "dico quod hic traditur modus operandi, sed modus operandi particularis propter convenienciam et disconvenienciam particularium oportet frequenter mutare": ibid., fol. 12v; also ibid., fol. 38r. Also see chapter 3, pp. 54-55 above.

[5] The ten *ingenia* or approaches are based on the disease (further subdivided into eleven considerations, such as the essence of the disease, pain, inflammation, physician and attendants, and efficacy of medications) and on the complexion, function, importance, anatomy, composition, connections, proximity, position, and sensitivity of the organs.

[6] "A medico necessitas observetur in omnibus iuxta naturam egritudinis et aliorum particularium": *De decem ingeniis*, fol. 283v.

[7] See pp. 132-133 above.

related to the various attitudes which for centuries slowed down the study of anatomy, especially the preoccupation with questions of natural philosophy and the aversion to the dissection of the human body, but it did not mean a total disregard of the need for anatomical knowledge. The preponderance of natural philosophy is illustrated in Bernard's *questio* whether the urine passes through the liver, as "some suppose": his negative answer is based on three arguments, first the distinction between "noble" and "base" organs, second his view of the circulation of the urine, and third his theory that the urine descends to the "chyle vein which is above the liver and near the kidneys and then is attracted" by the kidneys; "hence, those who believe that it passes through the liver speculate wrongly ... and such doubts do not occur except from defective [knowledge of] anatomy."[8] This conclusion is somewhat startling in view of the minimal use made in the argument of anatomical doctrine, let alone observation.

The observation of anatomy, together with its potential implications for practice, was mentioned more explicitly by Bernard in a few instances. He warned, for example, that for confirmed pleurisy one should phlebotomize the basilic vein on the afflicted side rather than in the opposite arm, and that "in this are mistaken all those who believe that the veins which nourish the ribs would comprise the vein in the arm, a thing that nowhere occurs to one who looks at (*inspicienti*) the anatomy."[9] That "inspection" apparently meant a combination of first-hand observation and consultation of the authors. Thus, where Bernard made a special effort to give detailed anatomical descriptions of intricate external organs such as the eye, ear, nose and throat, he drew his information primarily from Galen but he presented it in a way that suggests personal familiarity.[10] A similar

[8] "Queritur utrum urina que fuit missa ad totum corpus mittatur ad venas mediante epate quoniam aliqui opinantur quod mediante epate quoniam ipsum est medium inter venas tocius corporis et epar. Oppositum videtur quia membra vilia non mittunt ad nobilia sed epar est nobile. ... Dico quod urina que missa est ad totum corpus partim consumitur et partim in sudorem et vaporem convertitur ... alia autem pars postquam fecit iuvamentum venit usque retro ad venam kilim que est supra epar et prope renes, et tunc attrahitur; unde male ymaginantur qui credunt quod mittatur mediante epate quia ad epar numquam revertitur sed ad venam kilim, et iste dubitaciones non adveniunt nisi ex defectu anathomie": *De conservacione*, 2: *De urinis*, in Vatican MS Pal. 1174, fol. 37r.

[9] "Patet igitur [ei] qui secundum Avicennam volt anathomiam inspicere quod in pleuresi confirmata debet fieri flebotomia ex basilica eiusdem lateris.... In hoc autem sunt decepti omnes qui credunt quod vene que nutriunt costas constituant venam in brachio quod nusquam reperitur inspicienti anathomiam": *De regimine acutorum*, in Vatican MS Pal. 1083, fol. 280r.

[10] See *Lilium medicine*, fols. 37r, 41r, 43r and 48r respectively.

combination is reflected, with some paradox, in his discussion of epilepsy. He assertively claimed as his own ("*dico*") certain clinical observations and the causal explanation by humors that ascend to the brain; for the latter he used the argument that "it is evident from epileptic animals in which a horrible, fetid wateriness is found when they are dissected (*cum anathomatizantur*)."[11] However, the reader who would be misled into admiring Bernard's laboratory research may be disappointed upon finding the same statement verbatim in earlier works, for example in the *Compendium medicine* or *Laurea anglica* of Gilbertus Anglicus (ca. 1250).[12]

It is perhaps most surprising to learn that Bernard did not mention anatomy among the subjects to be mastered by the surgeon. Rather, he required that, like the physician, "the surgeon (*restaurator*) must be careful to know the complexion, sensitivity and nobility of the organs, and the nature of medications."[13] In general, he referred to a qualified surgeon as trained or learned, "*doctum cyrurgicum*."[14] The dual character of this learning, comprising both bookish and practical lore, is indicated in his advice that for ailments which were difficult to cure "we should seek recourse to a literate and experienced surgeon."[15] Such recourse would not be easy, however, for Bernard was openly sceptical about the availability of surgeons with theoretical and practical knowledge. At the end of his chapter on hemorrhoids he remarked that "the cause cannot be properly investigated except by surgeons, and they are mostly ignorant (*ignorantes*)."[16] In his checklist for the examination of an infected wound he instructed that one should see whether the complications resulted from

[11] "Dico quod humores evaporando possunt ascendere ad cerebrum et potissime illi ex quibus fit epilepsia cum sint venenosi horribiles et fetidi sicut patet in animalibus pacientibus morbum caducum cum anathomizantur in quibus invenitur aquositas horribilis fetida": ibid., fol. 34r.

[12] *Compendium medicine Gilberti Anglici...* (Lyon, 1510), fol. 109v. In this work, however, the etiology of epilepsy revolves around lunar influences, rather than around humors as in Bernard's *Lilium*.

[13] "Restaurator debet esse sollicitus in cognoscendo complexionem membrorum et sensum et nobilitatem, et naturam medicaminum, et ita poterit eligere ... et ita racio precedere debet": *Lilium medicine*, fol. 18rb; also ibid., fol. 18ra.

[14] Thus, for tumors on the coniunctiva "non est consilium corpore mundificato, nisi quod per doctum cyrurgicum extirpetur": ibid., fol. 38v.

[15] After several prescriptions for outgrowths on the coniunctiva Bernard added, "si autem ex hiis non curetur recurramus ad cyrurgicum literatum et expertum": ibid., fol. 38r.

[16] "Debemus pronosticare quod omnes [hemorrhoide] universaliter sunt difficilis curacionis quia pars illa est enormis et rugosa et non potest perscrutari causa sicut decet nisi fiat per restauratores, et illi ut plurimum sunt ignorantes": ibid., fols. 75v-76r.

the corruption of the humor or of the wound, or from the ignorance (*impericia*) of the surgeon."[17] Similarly, he warned that there could be no cure in certain cases where too much tissue had been "removed by an unlearned (*indoctum*) surgeon."[18] His strongest indictment was formulated in such claims as, "I have seen many people in jeopardy in the hands of surgeons,"[19] or more bluntly, "let us beware because I have seen many die in the hands of surgeons."[20]

With Bernard's vitally important requirement that the practitioner be "knowing" in matters of theory and practice we may consider as corrolaries several of the qualifications which he borrowed from Haly without further elaboration, specifically a good memory, intelligence and judgment. The ever growing amount of data and principles to be remembered by the physician was one of the principal stimuli not only for the condensation of practical compendia,[21] including the *Lilium medicine*,[22] but also for the logical arrangement of more specialized subject matter into "*tabule*" such as the work *De decem ingeniis curandorum morborum*.[23] The need to remember quickly certain rules may also account for some of the apparently mnemonic verses cited by Bernard. One, for example, indicated the diagnostic distinction between dyspnea and asthma;[24] another dictated which side of the body should be phlebotomized in each season;[25] and a third formulated the psychosomatic function of each major organ.[26] No learning by rote, of course, could make into a good physician anyone who lacked the intelligence to apply

[17] "... si proveniunt ista ex malicia humoris aut vulneris aut ex impericia restauratoris": ibid., fol. 17v.

[18] "Si autem caro sit diminuta et ablata per indoctum cyrurgicum ... non est curacio quia quod remotum radicale est non recipit curacionem": ibid., fol. 38v. Bernard also used the term "unlearned" when he cautioned that "obstetrix indocta" might be one of the causes of uterine abscess (ibid., fol. 93r) or of difficult childbirth (ibid., fol. 95v).

[19] Of the patients with hernia "multos vidi periclitari in manibus restauratorum quia perforabant intestina et stercus egrediebatur et moriebantur": ibid., fol. 91v.

[20] In case of abscess in the ear, "vigilemus igitur quia multos vidi mori in manibus restauratorum": ibid., fol. 42v.

[21] See Demaitre, "Scholasticism in Compendia," p. 85.

[22] "Quia igitur memoria hominis labilis, quod communiter de practica dicitur repetere humilibus non erubesco": *Lilium medicine*, prologue, fol. 1r.

[23] Bernard himself referred to this work as "libellus qui intitulatur tabula ingeniorum": ibid., fol. 99v.

[24] "Stat versus, Disnia se celat, canit asma morbumque revelat": ibid., fol. 55r.

[25] "Stant versus, Estas ver dextras [venas], autumpnus hyems oportet sinistras [incidere]," *De conservacione*, 1: *De flebotomia*, fol. 19r.

[26] "Cor sapit et pulmo loquitur, fel commovet iras; splen ridere facit, cogit amare iecur": *Lilium medicine*, fol. 83v. See also pp. 107-108 above.

that learning. This quality, which might perhaps be called plain common sense, was exemplified by Bernard in a critical comment on Heben Mesue's prescription of *repercussiva*, or medicines that reduce swelling, for the treatment of pleurisy. He argued that such medicines were effective only at the very onset of the disease, so that the practical value of Heben Mesue's prescription depended on several conditions. It would be necessary not only that the physician "happened to be right there" but also that "God had given him the great favor of seeing the disease coming on and of having the *repercussiva* in his hand or bag." But, Bernard continued, "it is not clear when all these conditions could coincide." Therefore, he concluded mockingly, "I am very surprised that the good Heben Mesue was dozing."[27] The irony of this application of common sense can be better appreciated in the light of Bernard's praise of Heben Mesue's *Antidotarius*, which he concluded (perhaps tongue-in-cheek) with the remark, "I understand that no one should dare to practice in good conscience without Heben."[28]

As may be gathered from the preceding example, a crucial function of the practitioner's intelligence and common sense was to make the right judgment about the exact nature and stage of the disease and about its appropriate treatment. With regard to the first aspect Bernard berated his peers by asserting that "there are few physicians who know well how to diagnose."[29] His own record, however, seems rather inconsistent. On the one hand, his diagnosis of epilepsy and especially of *petit mal* seizure is quite remarkable:

> I have often seen [the paroxysm] so short that the patient had only to lean against a wall or the like, rub his face, and it ceased. Sometimes he did not need a support, but there came to him a dizziness in the head, and blindness in the eyes, and he himself, sensing it, recited the Hail Mary, and before he had finished it, the paroxysm had passed.[30]

[27] "Si aliquo modo competerent [repercussiva] sicut voluit Ebemesue, tunc competerent in inicio iniciante, si quis presens esset, et Deus dedisset talem graciam quod principium inicians cognosceret, et quod portaret in manu aut in bursa medicamina repercussiva; sed quando possunt talia concurrere non videtur. Et ideo summe miror quomodo bonus Ebemesue solus dormitabat": *Lilium medicine*, fol 57v.

[28] "Quia ea que sunt in antidotario sunt optime composita ideo circa laudes non insisto et intelligo quod nullus audeat operari cum bona consciencia sine hebe, ideo de racionibus reddendis in composicione medicinarum pauca dicam quia ex illo optime potest extrahi": *De conservacione*, 2: *De urinis*, fol. 36r.

[29] "Pauci sunt medici qui sciant bene distinguere," *Lilium medicine*, fol. 52r. Also see p. 56 above.

[30] *Lilium medicine*, fol. 33r; tr. William G. Lennox, "Bernard de Gordon on Epilepsy," *Annals of Medical History*, 3 (1941), 374. See also p. 54 and n. 93 above.

On the other hand, we have seen earlier how he failed, to his own frustration and despair but nevertheless as the result of inadequate technique, in his attempts to diagnose leprosy.[31] Furthermore, it is worth noting that he seldom referred to manual exploration or palpation as diagnostic methods. He seemed to mention and to gloss over such methods rather casually, for instance when he stated that the diagnosis of wounds could be arrived at "through information from the patient or from the assistants, or through inspection by eye or by finger, and in similar manners which are contained more completely in the books of surgery."[32] He underscored the importance of questioning when he quoted Damascenus, "*above all* you should not be ashamed to interrogate the patient."[33]

Bernard's own favorite methods appear indeed to have been careful observation and interrogation. Of the clinical cases which he reported from personal experience and which included such earlier cited phenomena as *petit mal*, incomplete leprosy, and endemic leukemia,[34] none gives us a clue about a diagnosis by other means. This, of course, raises the question how much Bernard in his own practice relied on the two most celebrated methods, to which he devoted separate treatises, namely the examination of urines and pulses. In the absence of personal testimony, his use of uroscopy may perhaps be inferred from his statement that "the science of urines is quite familiar to us (*valde nobis domestica et familiaris*),"[35] or from his assurance that "through the urine we will be able to know the condition of health, disease, and neutrality. This makes evident the arrogance and ignorance of certain people who disdain and loathe the judgment of urines."[36] With regard to this "judgment" we are reminded of the difference between universal principles and particular circumstances, and of the intelligence needed to apply theory to practice: in spite of his conclusion to a lengthy *questio*, "that it does not suffice to have a vague and confused rather than a certain

[31] Pp. 124-125 above.

[32] "[Signa] ex noticia pacientis habentur, aut assistencium, aut cum oculo aut digito et similibus, que perfectius in libris cyrurgie continentur": *Lilium medicine*, fol. 17v.

[33] "Iuxta illud Damasceni, Et super omnia non pudeat te interrogare pacientem": ibid., fol. 53v.

[34] See p. 8 above.

[35] *De conservacione*, 2: *De urinis*, fol. 30v.

[36] "Per urinam igitur poterimus cognoscere disposicionem sanitatis, egritudinis et neutralitatis. Ex hiis igitur patet arrogancia et ignorancia quorumdam qui dedignantur et nauseant iudicia urinarum": ibid., fol. 28ra.

and definite knowledge (*scienciam*) about urines,"[37] Bernard admitted that "every judgment will be estimative."[38] A special demand on the physician's intelligence and perspicacity might be made by "deceivers" who sometimes brought animal urine "to fool the physician."[39]

Deception would not be a problem in the other diagnostic method, that of examination of pulses. This method, however, Bernard considered extremely difficult and to be mastered by "skill and long practice" rather than by theoretical instruction.[40] His own treatise on pulses was extremely brief and general, and his avowed purpose was to reduce their infinite complexity to a few simple guidelines.[41] An extant folio of an early French manuscript of the treatise portrays him in a beautifully illuminated initial as feeling a patient's pulse,[42] but this can hardly be considered prima facie evidence of his diagnostic procedures. The fact that these procedures included pulses may be inferred more plausibly from some of his own statements. He discussed, for example, the *questio* "whether a more certain diagnosis is obtained from the pulse or from the urine," and after elaborate arguments for the advantages of uroscopy he conceded that "about life, death, and the hour of death we can judge better by pulse."[43] Another suggestion that Bernard probably did not ignore pulse reading in his practice was implied when he compared the physician to a "good judge" whose sentence would be based on the testimony of "the pulse and urine and so on" or, even more elliptically, of "the pulse and other [symptoms]."[44]

[37] "Concluditur quod non sufficit habere scienciam de urinis vagam et confusam sed certam et determinatam": ibid., fol. 28rb.

[38] "Omne enim iudicium erit estimativum": ibid.

[39] "De urina autem quadrupedum non tractat medicus, sed deceptores apportant aliquando ut illudant medicum. Sed tamen Avicenna docet cautelam. .": ibid., fol. 29v. See further chapter 3, pp. 91-94 above for treatises on "*cautele*" against such deception.

[40] "De pulsibus dico tantum quod qui habebat artem et exercicium longum minus de pulsibus ignorabit": *Liber pronosticorum*, in Vatican MS Pal. 1174, fol. 91r. See also pp. 66-67 above.

[41] "Et quia ista sunt valde difficilia, ideo solus Deus et Galenus servus eius habuerunt noticiam huius pulsus ordinati. Nullus tamen desoletur propter ista, quoniam ars est finitum infinitatis compendium, et ideo tota ista multitudo reducitur ad tria...": *De conservacione*, 3: *De pulsibus*, fol. 52r.

[42] Bernardus de Gordonio, *De pulsibus*, single vellum folio, first half 14th century (with a counterpart, the first folio of his *Regimen sanitatis*, which portrays him while teaching) at the New York Academy of Medicine, Rare Book Room.

[43] "Dubitaret aliquis a quo accipitur cercius signum vel a pulsu vel ab urina.... Pulsus immediate attestatur supra disposicionem virtutis vitalis, et ideo de morte et vita et hora mortis melius scimus iudicare per pulsum quam per urinam": *De conservacione*, 2: *De urinis*, fol. 30r-v.

[44] "Iste autem medicus ut bonus iudex undique certificatur per naturam accidencium

With the diagnosis of fatal illnesses and the analogy between physician and judge we cross into another area of medical practice in which adequate knowledge and sound judgment would be crucial, namely prognostication. Bernard's analogy was derived from the etymological connection between *iudicium* and *crisis*,[45] and one of the most important keys to an accurate prognosis of diseases, especially of fevers, was the knowledge of their "critical days."[46] Hence his treatise (as well as that of Galen) on prognostication also bore the title *On Crisis and Critical Days*.[47] Since those critical days, on which the disease would take a turn for the better or for the worse, were supposedly influenced by the heavens, their computation was traditionally based on the movements of the celestial bodies. It was in this tradition that the knowledge of astrology became an essential requirement for the physician.[48] Whereas the consequences of seasonal changes accounted for the Hippocratic idea that "astronomy contributes not a little, but a very great deal, to medicine,"[49] the belief in specific influences of heavenly "complexions" was the primary reason why, as Bernard asserted, "the physician should know some things about the nature of heavenly bodies."[50] It may be noted that Bernard's requirement was less than categorical. A tinge of reservation was also apparent when he justified his discussion of astrology: "I have put down these few things for no other reason than because one is supposed to know according to his ability the nature of the heavenly bodies, for other-

---

si viderit quod pulsus et urina et ita de aliis testificantur bene pro natura.... Si autem pulsus et alia bene testificantur pro morbo medicus debet absolvere morbum et condempnare naturam et pronosticare ipsum moriturum": *Liber pronosticorum*, fol. 97v.

45 "Crisis sumpsit nomen a iudiciis.... Crisis enim interpretatur iudicium per hanc similitudinem: in vero enim et bono iudicio quatuor sunt necessaria ut actor, reus, testes, iudex...": ibid., fol. 97r.

46 For a survey of this doctrine see Karl Sudhoff, "Zur Geschichte der Lehre von der kritischen Tagen im Krankheitsverlaufe," *Archiv für Geschichte der Medizin*, 21 (1929), 1-22.

47 See chapter 3, pp. 41-43 and notes.

48 For Bernard's views on astrology and their implications for his own knowledge of the quadrivial art of astronomy see also chapter 1, pp. 20-22 above.

49 *On Airs, Waters, and Places*, tr. Adams, in *Theory and Practice*, ed. Kelly, p. 20. It is impossible here to deal with the relation between medicine and astrology beyond the views expressed by Bernard, although I hope in the near future to explore further his indebtedness to or differences with such sources as the pseudo-Ptolemaic *Centiloquium* and the pseudo-Hippocratic *Astronomia*, especially in the texts quoted in nn. 50-53 and 55-60 below.

50 "Si igitur ordo et regimen est in mundo et vultus inferior subiectus est vultui superiorum, tunc medicus debet scire aliqua de naturis superiorum": *Liber pronosticorum*, fol. 101r.

wise he would not be perfect in the science either of prediction or of treatment."[51] His introduction to the discussion was more ambiguous: "the physician, willingly or unwillingly, by hook or by crook (? *subponendo*) or I do not care in which way, is bound to know at least the nature and complexion of the signs and of the erratic stars."[52] After describing those, however, he concluded with greater moderation: "but because the physician can perhaps not know the equation of all the planets, in order that he may excuse himself he ought to have a table of the lunar equation ... and one accurate lunar calendar, for without these he is not a physician."[53]

To be sure, neither the limits which Bernard accepted in the physician's astrological proficiency nor the scepticism which he displayed towards some practical applications, as we will see later,[54] meant that he doubted or rejected the traditional belief in the influence of the stars on terrestrial life. He summarized with lucidity, though without evidence of personal adherence, this belief from such diverse sources as Augustinian doctrine, Neoplatonic writings such as the pseudo-Ptolemaic *Centiloquium*, and Galenic treatises. Thus he constructed the prologue to his *Liber de conservacione vite humane* around the paradigm of the three "custodians" of human life, namely "the spheres and stars, the divine intelligences, and God." Of these three, he explained, "God guards directly the human will, the angels the intellect, and the heavenly bodies the body."[55] He also cited the "teachings of the Egyptians" that the stars influence, from the moment of conception, not only man's destiny,[56] but even his very chance to be

[51] "Ista autem pauca posui non ad alia nisi quia quilibet tenetur scire suo posse naturam superiorum, aliter enim non poterit esse perfectus in sciencia predicendi nec eciam operandi": ibid.

[52] "Medicus velit nolit aut subponendo aut quoquomodo non curo tenetur ad minus scire naturam et complexionem signorum et naturam et complexionem stellarum erraticarum": ibid., fol. 100r.

[53] "Sed quia forte medicus adequacionem omnium planetarum scire non potest, saltem de aspectu lune per comparacionem ad signa non potest se excusare; ideo medicus tenetur habere tabulam adequacionis lune quoad medium motum, que communiter habetur, et tenetur habere verum et rectum calendarium lintoniensis [*sic*, lunacionis]: sine hiis autem non est medicus": ibid., fols. 100v-101r.

[54] See pp. 163-165 below.

[55] "Custodes autem tres sunt: orbes et sidera, et intelligencie divine, et Deus cuius nomen sit benedictum.... Cum ergo Deus immediate custodiat voluntatem et angeli intellectum et corpora celestia corpus humanum, patet quod magnificencia huius principatus [sc. hominis] est rutilans et preclara...": *De conservacione*, prologue, fols. 9v-10r.

[56] "Est intelligendum secundum doctrinam egypciorum quod secundum naturam constellacionis et influencie omnium orbium in hora casus seminis in matrice accidit

born alive.[57] He was careful, nevertheless, to avoid potential charges of determinism, and he pointed out that "the spheres do not influence with necessity, but in the way of inclination and disposition so that someone, who is inclined by the stars towards one thing, will never be equally disposed towards its opposite."[58] Bringing these philosophical principles closer to medical theory, he cited Galen's view that there are two foundations or "roots" in every development: the *radix superior* and, influenced by it, the *radix inferior*. The former, which "adheres to one way and order" as do "the movements of the spheres," is the source of "all that is beautiful and honorable"; the latter, whose course is erratic and variable, "is the principle of disease."[59] Relating this view to prognostication Bernard specified that the course of chronic diseases is controlled by the sun but that of acute diseases by the moon.[60] Furthermore, the sun rules the heart and the "vital force" whereas the moon governs the "natural force" and thereby the *crisis*, so that the physician "is not concerned with the solar month" but he "is bound to measure the lunar month" in predicting the future course of a disease.[61]

---

membrorum lineacio et figura, et secundum naturam influencie hora nativitatis regulabitur in tota vita in moribus, et a vultu cum vultus inferior sit subiectus vultui superiorum": *Liber pronosticorum*, fol. 100r. See also notes 57-59 below.

[57] "Natus in mense vii. bene potest vivere quia septem planete iam ibi operati sunt, in viii. autem non quia debilis videtur postquam in vii. non oritur, quia oportet expectare usque ad ix. et potissime quia iterum incipit dominare Saturnus in viii., sicut fecit in prima, et quia Saturnus immobilitat et inducit quietem cum sit planeta frigidus et ideo si movetur contra inclinacionem orbis movetur": *Lilium medicine*, fol. 96r.

[58] "Cum orbes non influant secundum necessitatem, influunt tamen secundum aptitudinem et disposicionem, ita quod si ad hoc aliquis [sit] dispositus quantum est ex orbe, numquam erit in contrario ita aptus": *Liber pronosticorum*, fol. 100r.

[59] "... figurabatur in dictis Galeni iii. de Creticis: quedam est radix superior et ista adheret vie uni et ordini uni sicut sunt motus orbium; et quedam est radix inferior et ista non adheret vie uni nec ordini uni et est principium morbi": *De conservacione*, 4: *Regimen sanitatis* prologue, fol. 53v. Also, "est intelligendum quod natura dierum criticorum sumitur a simplici radice [*sic*, a duplici indicacione], quarum una est inferior sumpta a natura morbi, que quidem est agitabilis et variabilis, non adherens vie uni nec ordini uni; alia autem est sumpta a radice superiori, que quidem non est agitabilis nec variabilis, adherens vie uni et ordini uni: unde omne quod est pulcrum et honestum de sursum provenit": *Liber pronosticorum*, fol. 99v.

[60] "Chronice [egritudines] autem reguntur per motum solis cum sol non peraget [*sic*, peragret] zodiacum nisi in anno et ideo quia movetur motu tardo.... Acute autem regulantur per motum lune, cum luna peragret totum zodiacum in mense uno": ibid.

[61] "Nunc autem sol habet aspectum ad cor et ad virtutem vitalem, et quia nos loquimur de mense per comparacionem ad crisim et crisis non fit per virtutem vitalem et per consequens nec per virtutem solis, ideo de mense solari nihil curat medicus. Nunc autem luna habet aspectum ad virtutem naturalem et quia crisis fit per virtutem naturalem, ideo medicus tenetur considerare mensem lunarem": ibid., fol. 101r.

In a condensed paraphrase of the well-known introduction to the Hippocratic *Book of Prognostics* Bernard outlined the important advantages of an accurate prognosis:

> The knowledge of prediction is useful for many things because, having heard a correct prognostication, the patient is made obedient and feels that the physician is able to treat him. It also prevents future dangers and saves the physician from subsequent infamy and dishonor when he knows the condition and knows how to prescribe the regimen and diet, to administer syrup and medication, and to apply particular remedies.[62]

After having diagnosed and prognosticated, the physician would indeed have to draw even more upon his knowledge and judgment in choosing for his therapy one or several of the following methods: diet and regimen, medication, and surgery. This was the order of preference, and Bernard insisted that "a better physician is he who is able or knows how to cure diseases with food and drink" rather than with other means.[63] In contrast with the unqualified recommendations of more drastic methods in, for example, the *Regimen sanitatis Salernitanum*,[64] he cautioned that "phlebotomy and drugs should be used with great reserve lest the strength be weakened, because it is the strength that cures."[65] He called the conservation of strength (*virtus*) "the mistress of all the methods of treating diseases."[66] He further believed that the physician, as the "servant of nature," should "stimulate nature" rather than work against it.[67]

[62] "Sciencia autem perficiendi predicendi [*sic*] est utilis ad multa, quia audita pronosticacione vera, paciens redditur obediens et ymaginatur quod medicus est potens ipsum curare. Subvenit eciam futuris periculis et preservat medicum ab infamia et dehonestate, et per consequens postquam cognoscit statum [sciet] et dietare, cibare, syrupum et medicamen applicare, et particularia remedia adhibere": ibid., fol. 73v. Compare the Hippocratic *Liber pronosticorum* in a Latin translation that would have been familiar to Bernard, for example in the *Articella* (Venice, 1487), fol. 47r, at the New York Academy of Medicine.

[63] "Medicus igitur melior est, qui egritudinem cum cibis et potibus curare poterit vel sciverit": *De decem ingeniis*, fol. 283v.

[64] See *The Regimen of Health of the Medical School of Salerno*, ed. and tr. Pascal Parente (New York, 1967), pp. 58-59.

[65] "Flebotomia et farmacia sunt multum conservanda ne virtus debilitaretur, quia virtus est que curat morbos": *Lilium medicine*, fol. 56v. This caution may indicate a difference with Arnold of Villanova, who used phlebotomy as the typical treatment in his work *De considerationibus operis medicine* in *Opera* (Lyon, 1509), fols. 90v-101v; I am preparing an annotated edition of this work.

[66] "Intendamus igitur in custodia[m] virtutis tamquam magistram omnium ingeniorum curandorum morborum": *De decem ingeniis*, fol. 284r.

[67] "Excitanda igitur est natura cum medicus sit minister eius": *Lilium medicine*, fol. 6r. This idea is further examined in my paper, "Nature and the Art of Medicine in the Later Middle Ages," *Mediaevalia*, 2 (1976), 23-47.

While one may be inclined to think of surgery as "against nature" it is worth noting that the term "restorer (*restaurator*)" was used frequently by Bernard for the surgeon.[68] Still, in the series of treatments from which the physician must choose, he did rank surgical procedures as the last resort. It was clearly his custom to leave the patient to the hands of the surgeon only when a disease had become "difficult or almost impossible to cure" by the physician,[69] or when the treatment was "difficult, dangerous and lengthy, and required a major procedure."[70] Thus he advised that for a patient with scrophula, after all other methods had been exhausted, "it is necessary to have final recourse to the hand of the surgeon, who should watch out for the veins and arteries; and if he does not succeed, let us go to the kings."[71] Those patients who were afraid of the knife could be treated with "cautery and with goat's tallow and vinegar," but Bernard added, "because this treatment is unsightly and difficult I leave it now completely to the surgeons."[72] Also the treatment of bloody ulcers he would "yield to the surgeon who will cure it with a hot iron (*cum ferro et realgar*), God willing."[73] In addition to the treatment of wounds and fractures, the most drastic and unsightly procedures, and cautery, phlebotomy was apparently also the function of the surgeon.[74] Bernard

[68] See, for example, nn. 13, 16-17, and 19-20 above and 69-72 below. Professor Richard Lemay has pointed out to me that the term was obviously a translation of the Arabic *mujabbir* or *jābir* (from "to reset") which was used, for example, by Albumasar (Abū Ma'shar, d. 885) in the *Liber introductorius*. Albumasar, however, applied the term exclusively to the "bonesetter," whereas Bernard clearly used it for the surgeon in general. I have thus far not found the term in other Latin medical writings. It would be the exact synonym of *cyrurgicus* according to the 1498 incunabulum of the *Lilium medicine*, fol. 21r, where we read "regimen medicorum et restauratorum, scilicet cyrurgicorum." London BL MS Harl. 3698, fol. 20v, however, reads "*et*" rather than "*scilicet*."

[69] For example, because gangrenes "sunt difficiles et quasi impossibilis curacionis, ideo curam restauratori relinquamus" (ibid., fol. 10v); the same for fistula, cataract, hemorrhoids, *noli me tangere*, and so on (ibid., passim).

[70] Treatment of a fractured skull "est difficile et periculosum et prolixum et maiori solemnitate indiget, ideo restauratoribus derelinquo": ibid., fol. 18v.

[71] "Finaliter oportet recurrere ad manum cyrurgicam, sed caveat sibi restaurator a venis et arteriis et si non vadamus ad reges": ibid., fol. 12v. On the "royal touch" see further chapter 1, p. 3 and n. 10.

[72] "Aliqui sunt timidi et timent incisionem ... in talibus rumpatur locus cum cauteridicis et sepe ircino et aceto; sed quia ista cura est turpis et difficilis nunc totaliter restauratoribus derelinquo": *Lilium medicine*, fol. 12v.

[73] "Ulcera saniosa et antiqua sunt difficilis curacionis, et ideo curam cyrurgico relinquamus qui cum ferro et realgar represso curabit si deus voluerit": ibid., fol. 10v.

[74] "[Flebotomia] cum lancea que fit per manum restauratoris": *De conservacione*, 1: *De flebotomia*, fol. 10v. Also, for abscesses of the uvula, "restaurator percutiat uvulam cum flebotomo": *Lilium medicine*, fol. 49r.

provided evidence, however, that venesection by others was not unusual. In one case he recommended a certain prescription as "useful for barbers so that they may be able to strike the center of the vein."[75] In another, perhaps more surprising reference he reported his personal performance of venesection in a manner that suggests nothing extraordinary except the fact that he had ignored an adverse astrological consideration.[76]

Even more fascinating are the indications that Bernard, although he relegated the manual operations to the surgeons and referred his readers to "the books of surgery" for the details or "length and width" of wounds and fractures,[77] was himself quite familiar with several surgical procedures and may have applied them personally. Those indications seem implicit in certain details, for example his insertion of "for the moment" when he left aside a discussion of surgical methods.[78] Suggestive, too, are the instances in which he used the phrase "let us have recourse to the surgical hand (*ad manum cyrurgicum*)" rather than "to the surgeon," and then proceeded to describe the operation in some detail.[79] Of special interest in this regard are his descriptions of paracentesis or tapping for peritoneal dropsy,[80] of the incision of breast cancer,[81] and of

[75] "[Medicamen] valet barbitonsoribus, ut possint percutere in medio vene": ibid., fol. 98r-v. It may be noted that Bernard's references to barbers are few in number despite his elaborate discussions of phlebotomy, and they are implicitly unfavorable (see n. 151 below).

[76] See p. 164 below.

[77] "Si vulnus fuerit ex causis extrinsecis, ex noticia pacientis habentur ... aut cum oculo aut digito et similibus que perfectius in libris cyrurgie continentur": ibid., fol. 17v. "Diversa sunt vulnera et contusiones et fracture et lesiones diverse valde, ideo illam latam longitudinem restauratoribus derelinquo": ibid., fol. 25v.

[78] "Curam autem lapidis [renum] factam manu cyrurgica *quantum est de presenti* penitus derelinquo": ibid., fol. 87r. Also note the "*nunc*" in n. 72 above.

[79] Thus, if nothing else cures an abscess on the eylid, "ad manum cyrurgicam recurramus et interior palpebrarum inversatur et abscindatur eciam nodus ... abstrahatur, et curetur vulnus sicut vulnera debent curari," and Bernard further described in detail several other eye operations (*Lilium medicine*, fol. 40v).

[80] "Ultimo autem et in desperacione et cum non videmus aliam medicinam inscidatur per duos digitos subtus umbilicum et fiat foramen dyagonale ita ut pellicula exterior inscisa cum venimus ad cyphac inscidatur cyphac subtus primum vulnus et cutis exterior cooperiat foramen chiphac ut aqua subito exire non possit et tunc imponitur canalis per foramen primum et postea inclinetur capud canalis et intromittatur per secundum foramen et tunc aqua extrahatur paulatine": ibid., fol 81r.

[81] "Si ergo locus sit talis quod possit incidi sicut in mamilla et similibus incidatur locus totus omnino cauterizatur et permittatur fluere multus sanguis, hoc autem fiat cum magna cautela quoniam periculosum est; apponatur ergo medicina sine mordicacione. Rx... sed quia passio ista est difficilis curacionis aut impossibile postquam antiquata est, ideo deinceps restauratoribus relinquamus": ibid., fol. 11v.

the suturing of a wound. Suffice it to quote an excerpt from his instructions on how to "begin, with God's help, operating on a wound that is large." After the application of a styptic,

> let us join the lips of the wound while being careful that there be neither oil nor dust, nor any foreign matter in the wound. Next let us sew – the thread should be of silk, smooth and even, and the needle should be of suitable size – and in each point there should be a knot until the end; and then let us make a tie. Folded gauze (? *plumaceoli duplicati*) is then put over it, and care should be given that no water come close to the wounds....[82]

The suggestions and descriptions cited here deserve to be weighed against Bernard's references to surgery as extraneous to his medical teaching and practice. Especially his familiarity with details seems to bear out the view of several historians that at Montpellier surgery was taught with medicine and practised by physicians.[83] It is also possible that, at least in serious cases, physicians were expected to supervise a surgeon's operation, since Bernard reported the fatally excessive bloodletting by a surgeon "while no master was present." In this incident, however, Bernard may have been particularly incensed because the patient was his colleague or teacher and the operation had been agreed upon by a whole group ("*turba*") of masters; therefore, he was perhaps not insinuating the violation of a norm but rather censuring the physicians for their failure to stay with the case, to exhibit due concern, or to apply adequate diligence.[84]

When he required diligence as an essential quality of the practitioner, Bernard expressed his belief that even the most "knowing" physician would be found wanting if he neglected to apply his theoretical and practical knowledge to the care of health and life. The vital interdepen-

[82] "Cum dei auxilio sic incipiamus operari vulnus quod est magnum ... deinde coniungamus labia vulneris cavendo quod in vulnere non sit oleum vel pulvis nec aliqua res preter naturam. Deinde suamus et filum sit planum et equale de serico et acus quadrata et in quolibet puncto fiat nodus et insuper nodando in quolibet puncto usque ad finem. Deinde ligemus et plumaceoli [?] duplicati desuper apponuntur et cavendum ne aqua approximetur vulneribus...": ibid., fol. 18r.

[83] Thus, Rashdall, *Universities*, 2: 136, n. 2, adding that this "shows the superior position of the art as compared with the position it held at Paris." See also Sarton, *Introduction*, 3: 272; and Émile Forgue, *Sept siècles de chirurgie à la Faculté de Médecine de Montpellier* (Montpellier, 1920), p. 6. No new light is shed on this question in the recent work of Louis Dulieu, *La chirurgie à Montpellier de ses origines au début du* XIX*e siècle* (Avignon, 1975).

[84] "Fuit mortuus magister noster ex litargia ... et erat ibi turba magistrorum, et omnes convenerunt quod aperirentur emorroydes; et cyrurgicus nullo magistro existente aperuit, et extraxit ut dicitur sanguinem in nimia quantitate, intantum quod semper processit debilitando donec mortuus est": *Lilium medicine*, fol. 27v.

dence between knowledge, practice, and diligence was underscored especially in the *modus operandi* laid down by Bernard for baffling cases in which the medication had no apparent effect. He advised the physician, on the one hand, who knew in the process whether or not he had taken the right course, either to continue the medication or "not to persist in his error but to correct himself."[85] Bernard's statement elsewhere, that in the regular administration of the patient's diet "it is necessary every day to correct errors,"[86] further stressed the dynamic aspect of a continuously adjusted knowledge. If the physician was, on the other hand, "entirely ignorant of what should be done," Bernard urged him to "desist because then nature will cure or make things clear, as Avicenna says." Since such inaction might prove embarrassing, Bernard suggested that the physician administer placebos, "which can neither help nor hurt"; but meanwhile he should zealously "bend over the books day and night" and re-examine the symptoms, anatomy and physiology, and therapeutics: "it is impossible that he who diligently considers these things will not know either the truth or an approximation of the truth, unless he be a millstone – and for such people I do not write."[87] The close correlation between knowledge and care was epitomized most succinctly when Bernard asserted that some difficult cases "cannot be known without study, concern, and frequent visits with the sick."[88]

For the historian who, influenced by several manuscript illuminations, visualizes the scholastic physician as solemnly enthroned in his cathedra, as well as for the person who nostalgically regrets the recent demise of the

[85] "Si enim effectum medicaminis non videris et in processu cognoscis quod non bene operaris nullo modo moreris in errore sed corrige teipsum; si autem recte operaris non mutes medicamen sed muta modum quia a consueto non fit alia curacio vel passio (non fit passio in aliis curacio: C), nam durum lapidem gutta cadendo cavat": *De decem ingeniis*, fol. 283v.

[86] "In regimine sanitatis ... quotidie oportet corrigere errores": *Lilium medicine*, fol. 66r. The incunabulum (Venice, 1498), fol. 67r cites Galen "libro suo de regimine sanitatis."

[87] "Si autem quicquid agendum omnino ignores dimitte dicit Avicenna quod tunc natura curabit aut manifestabit. Si autem timeas ignominiam, des ea que non possunt iuvare nec nocere, de quo sit curandum, sicut est simplex iulep et similia, et interim insiste libris die ac nocte, et considera colorem et tumorem et lesionem et operacionem et anatomiam, et ea que egrediuntur a corpore, et iuvancia et ledencia, et qui ad ista diligenter consideraverit non est possibile quin cognoscat aut secundum veritatem aut coniecturam proximam veritati, et si non est lapis molendini, et talibus non scribo": *De decem ingeniis*, fol. 283v.

[88] When the symptoms of choleric fever are difficult to read and the paroxysm is present, "non possunt ista sciri sine studio et sollicitudine et frequenti infirmorum visitacione": *De conservacione*, 2: *De urinis*, fol. 41v.

general practitioner and his daily rounds, Bernard's concept of medical care may be refreshing. Whether in his personal reminiscences of particular cases or in his general guidelines for the physician's practice, he consistently referred not to office visits but to house calls. Thus, when he elaborated upon his requirement of diligence, he explained that the practitioner should first

> visit the patient in the morning, especially when he is in danger and primarily for the sake of [examining] the urine and its residue so that he may see its color and substance before it becomes spoiled. He should also give instructions with regard to the patient's food and for other particulars, and he should prevent future dangers and keep in check the preceding ones.[89]

The anticipation and prevention of future dangers might not only affect the patient's recovery but also redound to the physician's credit. By his accurate prognosis the physician would inspire confidence,[90] "not degrade himself by vain and false promises,"[91] and escape "infamy and dishonor."[92] If, on the contrary, he failed to predict possible complications, he might be blamed for the patient's deterioration or even for his death.[93]

Bernard's advice to the physician who foresaw a fatal outcome seemed ambiguous. He followed the Hippocratic recommendation that, if the hopes of recovery were small, "one should try to escape from such cases, provided one can do so honorably."[94] Perhaps overstating that recommendation he urged that, when death appeared imminent, the

[89] "Secundo quod sit diligens, scilicet quod de mane visitet pacientem (pacientes: C), potissime quando est (sunt: C) in periculo, et potissime propter urinam et residenciam, ut videat et colorem et substanciam antequam corrumpatur, et quod ordinet de cibo infirmi et aliis particularibus, et quod obviet periculis futuris et reprimat precedencia": *De decem ingeniis*, fol. 283v.

[90] "Medicus igitur ut de eo fiducia habeatur, dicat aliquando aliquid de preteritis, aliquando de presentibus, aliquando de futuris": *Lilium medicine*, fol. 4v.

[91] "Nolite dehonestare vos ipsos cum vanis et falsis promissionibus": ibid., fol. 33r. Also, "utile est pronosticare ne impossibile promittamus" (ibid., fol. 21r); and "caveat sibi medicus de vanis promissionibus" (ibid., fol. 55r).

[92] "Sciencia predicendi ... preservat medicum ab infamia et dehonestate": *Liber pronosticorum*, fol. 73v (see p. 148 and n. 62 above).

[93] "Tunc dicent assistentes quod vere flebotomia [pacientem] interfecerit, aut postquam sumpserit syrupum semper est deterioratus, et ita [medicus] factus erit ignominiosus. Sed si pericula in quolibet morbo predixerit semper medicus commendabitur et nihil sibi imputabitur": *Lilium medicine*, fol. 2v.

[94] *On Fractures*, tr. Adams, in *The Theory and Practice*, ed. Kelly, p. 198. Bernard warns that in certain cases the symptoms indicate "mortem in ianuis, [et] in talibus non competit medicacio sed fugere quantum est possibile": *Liber pronosticorum*, fol. 88v; also ibid., fol. 94v.

physician should flee "hurriedly" or "at once, without greeting his host."[95] Paradoxically, he was nevertheless concerned that every attempt be made to postpone the inevitable for terminally or critically ill patients. For example, while assuming that advanced leprosy "will not be cured," he believed that "we can prolong life, and with medication prevent the poisonous melancholic matter from reaching the heart."[96] He also applied this concern to patients who were "approaching death: even then it is good to prolong life as much as possible."[97] For patients who seemed already dead because of apoplexy he first described several diagnostic tests, ranging from conventional means to discern breathing to drastic measures such as shouting the patient's name, pulling his hair, or pricking him with a needle.[98] All this may have been of little use because, in Bernard's words, "I say that I have seen many apoplectics and in truth I have never seen anyone escape."[99] Still, he listed various prescriptions which should be administered to the patient "without interruption until he revives or dies, so that it can not be said that he died from lack of medication."[100]

Lest the physician be found negligent in less dramatic cases, Bernard required as a matter of course that he ought to visit his patients a second time at noon. The purpose of this visit was purely diagnostic, "so that he may clearly see the stage, development, nature and composition of the diseases and their symptoms."[101] We begin to wonder whether the

[95] In case of *frenesis* "iam mors sedet in pedibus. Medicus igitur confestim fugere debet": *Lilium medicine*, fol. 32r. And if a comatose patient does not respond to stimulants, "Statim fugiat medicus hospite insalutato": ibid., fol. 58v.

[96] "[Paciens] non curabitur, sed vitam prolongare possumus et impedire cum medicaminibus quin materia melancolica venenosa veniat ad cor": ibid., fol. 13v.

[97] "Licet approximentur morti, tamen bonum est vitam prolongare quantum est possibile": ibid., fol. 73r.

[98] "Quando tu es vocatus ad pacientem et vides eum sine sensu et motu ... tunc voca cum proprio nomine et alta voce et trahe eum per capillos et conduplica digitos ... et punge ipsum cum acu et stilo.... Aliquando tamen est magnum dubium utrum sit vivus vel mortuus propter privacionem sensus et motus et hanelitus et pulsus et tunc claudantur omnia foramina domus et flosculum lane minutissime carpinatum ori eius apponatur et si moventur pili lane procul dubio adhuc vivit...": ibid., fol. 34v.

[99] "Dico ego quod ego vidi multos apoplecticos et in veritate numquam vidi aliquem evadere": ibid.

[100] "Curacio. Rx... et ista omnia per ordinem sine intermissione interim continuantur donec revixerit aut mortuus fuerit ut non possit dici quod ex defectu medicaminum deficiat": ibid.

[101] "Secundo debet visitare in meridie, ut possit bene videre terminum, actum, naturam, et composiciones morborum et accidencia eorum (videre omnium accidencium naturam et composiciones morborum: C)": *De decem ingeniis*, fol. 283v.

physician was here given a *modus operandi* that was either quite idealized or based on the presumption of a limited clientele,[102] when he was expected to pay yet a third call "towards night, so that he may arrange for sleeping remedies and medications, and for food for the following day." At the same time he should see to "all the other things that can be considered with regard to the patient, so that he may render him obedient and the assistants careful, and do everything essential for the patient's benefit."[103]

The considerations evidently included not only the patient's condition but also his physical and human environment, in addition to the particulars of the disease itself. If he lived in a hilly part of the city, he should be moved down – with caution so that, "while being carried, he be protected from strain, heat and cold." He should also avoid the irritating vicinity of workshops that caused air and noise pollution, such as those of "blacksmiths, hammermen, tanners, smelters of tallow, or men who work with sulphur."[104] Besides such "industrial" causes, urban causes of air pollution were not entirely unknown in the Middle Ages, as we learn from Bernard's observation that the air in certain large cities was "foggy and humid"; we may not find much help, however, in the "artifice" he recommended for that problem: "he who is unable to move to the mountains may purify the air by adding aromatics by means of fans."[105] With regard to other qualities of the air, the patient's house and room were also to be inspected for draftiness and humidity.[106] In short, everything around him should contribute both to "the greater purity of the air and to his delight,"[107] and should be "in accordance with his

[102] See, however, p. 156 below.

[103] "Tercio circa noctem ut ordinet sompnifera et medicamina et cibum in crastinum, et alia omnia que possunt considerari circa infirmum, quod reddat ipsum obedientem et assistentes diligentes, et omnia que intrinsecus sunt ad utilitatem infirmi": *De decem ingeniis*, fol. 283v.

[104] "Circumspicere debet igitur medicus disposicionem civitatis et domus, et si fuerit in parte decliviore civitatis quod faciat ipsum mutari a superiore parte ville cum cautela quod sit cautus in portando a labore et calore et frigore. Et debet advertere medicus si est officium iuxta pacientem contristans ipsum sicut sunt fabres et malleatores et paratores coriorum et fundentes sepum et qui operantur de sulphure": ibid., fol. 283r.

[105] "Aer eciam sit mundissimus et si non habeatur naturaliter habeatur per artificium, potissime in magnis civitatibus ubi aer est caliginosus et humidus... [see chapter 1, n. 30 above]; qui igitur non est potens se transferre ad montes, cum flabellis rectificet aerem aromatica apponendo": *Lilium medicine*, fol. 66r.

[106] "Et quod videat cameram si est iuxta terram et quod non sit reumatica, et si superius quod non sit nimis ventosa...": *De decem ingeniis*, fol. 283r.

[107] "Quod omnia declinant ad delectacionem et ad maiorem aeris puritatem": ibid.

wishes."[108] This rule applied, at least in part, to the patient's human environment. In order that his "regimen be conducive to gladness and joy, he should be in the company of those he loves"; but he should also be surrounded by "those whom he fears, so that they may keep him away from harmful things."[109]

When it came to the treatment itself, it was imperative that "the patient be obedient, because if he should refuse the [prescribed] phlebotomy, medication and local remedies, the disease would be prolonged or dangerous complications would ensue."[110] It was further important that the prescriptions be adjusted to the patient's physique: stronger medications for those who were more muscular because of habitual hard work, "such as sailors, men of arms, farmers and shepherds"; weaker medications for those with "delicate flesh like that of women," more specifically "they who are used to being in the shade and at rest, such as scholars, students, the religious, merchants, or others such as weavers, tapestry makers (or carpenters) and the like, who sit in their workshops all day."[111] This list of occupations is noteworthy because it indicates that Bernard did not think exclusively or even primarily of the rich or the noble as patients of the university trained physicians for whom he wrote down his principles of practice. He seemed, nevertheless, to assume that those patients would have some social standing, since he also formulated requirements for the domestic service and for the kitchen. For example, the cook should be "temperate," if not in appetite at least in constitution, "because if he should be phlegmatic he would judge something quite salty to be insipid."[112] As to the domestic servants, they should be "good, few, and diligent,"[113] although we are reminded that good help is hard to

[108] "Quod omnia que extrinseca sunt ... sint ad velle infirmi": *Liber pronosticorum*, fol. 98r.

[109] "Ingenium sumitur ex accidentibus anime scilicet quod regimen declinet ad leticiam et gaudium et moretur cum dilectis et cum illis quos timet ut prohibeant ipsum a nocivis": *De decem ingeniis*, fol. 283r.

[110] "Quod paciens sit obediens, si enim recuset flebotomiam, farmaciam et localia remedia, prolongabitur morbus aut venient periculosa sinthomata": *Liber pronosticorum*, fol. 98r.

[111] Among the stronger people "sunt marinarii et homines armorum et agricole et pastores et consimiles. In quibus autem tota caro est rara et mollis, sicut mulierum, illis debilis medicina competit, sicut omnium eorum qui consueverunt esse in umbra et quiete, sicut sunt scolares, studentes et religiosi et mercatores, quidam sunt draperii et carpentarii (trapesatarii: P) et consimiles qui quotidie sedent in operatoriis": *De decem ingeniis*, fol. 285r.

[112] "Quod coquus sit temperatus, si enim esset flegmaticus res multum salsa iudicaretur ab eo insipida": ibid., fol. 283v.

[113] "De ministris, quod sint boni et pauci et diligentes": ibid.

find.[114] Their main concern was to be for the patient's welfare and therefore "they should not bring him any bad news, and allow only a few people to come in." They were also to protect him not only from accidents, such as falls and suffocation, but also from irritations such as "the barking of dogs, lest his sleep be disturbed."[115] This cause of insomnia must have been rather common, since Bernard also mentioned it elsewhere together with "the abuse of bandits or the noise of drunks"[116] – perhaps another indication that he was thinking of the town rather than the castle as the milieu in which the academic physician commonly practised his profession.

Bernard's third requirement was that "the physician should practise according to the art, that is by the canons of Hippocrates, Galen, and the other sages." He warned that by this requirement were condemned all the arts of "soothsaying, prophecy and divination," as well as those of "suspending herbs around the neck, and all quackeries (*empirica*), sorceries, incantations, and numerous other things which are better not revealed on account of those who abuse them and who, neglecting their conscience, resort to magic."[117] This solemn exorcism seems to contradict several instances in which Bernard himself suggested the use of quack remedies and magical cures. For example, he recommended that herbs, "collected with the Lord's prayer," be suspended around the neck of a patient suffering a fistula,[118] or on the hip of a woman in difficult childbirth.[119] And in his most flagrant lapse into quackery, in the *Lilium*'s

[114] "Quod ministri sint diligentes, licet raro inveniantur": *Liber pronosticorum*, fol. 98r.

[115] "Et quod appetant salutem pacientis, et quod non nuncient ei aliqua sinistra, et non permittant intrare nisi paucos, et quod diligencia ministrorum advertat ad omnia que sunt intrinsecus [*sic*], scilicet ad latratum canum ne impediatur sompnus, et casum et ruinam; et si veniat sudor quod non sit discoopertus nec nimis cohopertus, et potissime quod non coarcent ipsum quia multi ex hoc suffocantur...": *De decem ingeniis*, fol. 283v.

[116] "Paciens debet dormire et quiescere sed angustiatur propter latratum canum aut insultum latronum aut matos rumores": *Liber pronosticorum*, fol. 98r.

[117] "Tercio oportet quod medicus operetur secundum artem, scilicet per canones Ypocratis et Galeni et aliorum sapiencium, et per hoc condempnatur omnis ars augurandi, vaticinandi et divinandi sicut est ars geometrialis (geometrica: C, P), et suspendendo (suspendendi: C, P) herbas ad collum et omnia empirica et sortina (sorticinia: C, sorticina: P) et fassina, et alia quamplurima que non est bonum revelare propter abutentes qui consciencia neglecta utuntur magicis et ingeniis (quibusdam: C, P) fangosis (?; fatigosis: C; fatuis: P)...": *De decem ingeniis*, fol. 283v. See also chapter 4, p. 129 above.

[118] "Scolopendria collecta cum dominica oracione et portata ad collum curat fistulam": *Lilium medicine*, fol. 11r.

[119] "Dicitur quod si colligatur agrimonia cum oracione dominica et ligetur in coxa mulieris laborantis in partu ... statim expellitur fetus": ibid., fol. 96r.

chapter on epilepsy, he prescribed as "frequently proven (*probatum frequenter*)" a truly magical remedy in which, appropriately enough, the Magi figured prominently. In this instance he gave assurance that the patient would be cured by hearing, spoken directly into his ear, the following verses: "Gaspar bears the myrrh, Melchior the frankincense, Balthasar the gold; who bears with him these names of the three kings is, God willing, delivered from falling sickness."[120] To this remedy, which was part of a respected tradition,[121] Bernard added others that were believed to help the epileptic patient. Most of these were presumably based on the occult or sympathetic qualities of such ingredients as mistletoe, wolf's hide, weasel's blood, cuckoo's ashes, or hare's curd,[122] and thus they might fall either into the category of *empirica* or into that of *experimenta* which we have examined in the preceding chapter.[123] Two more remedies, however, depended on incantations. That is, around the patient's neck were to be hung the verse about the three kings, cited above, and a Gospel passage that had been read by a priest over the patient's head.[124]

The seeming contradiction between Bernard's condemnation and recommendation of magical cures may in part be resolved by a closer

[120] "Cum aliquis est in paroxismo si aliquis ponat os supra aurem pacientis et dicat tres istos versus procul dubio statim surget: Gaspar fert mirram, thus Melchior, Balthasar aurum. Hec tria qui secum portaverit nomina regum solvitur a morbo Domino prebente (domini pietate Paris BN 16189) caduco ... probatum est frequenter quod statim surgit," ibid., fol. 33r.

[121] For strikingly similar remedies see Karl Sudhoff, "Magische Fallsuchtmittel," *Archiv für Geschichte der Medizin*, 2 (1909), 383-384. The formula was also recommended by, among others, Petrus Hispanus (ca. 1220-1277) and Peter of Abano (ca. 1250–1316): see Thorndike, *Magic*, 2: 497 and 904.

[122] "Multa alia sunt que curent epilensiam, ita ut viscus quercinus si portetur ad collum aut quocumque modo sumatur ... corrigia facta de pelle lupi si portetur iuxta carnem.... Si sanguis mustele misceatur cum urina et potetur multum confert.... Si catulus (cuculus: Venice, 1498, fol. 119v) ponitur in olla in furno ... et quod pulverizatur et pulvis epileptico exibeatur curat. Coagulum leporis multum valet...": *Lilium medicine*, fol. 33v. For an English translation (with some errors, such as "curds of a leopard" for *coagulum leporis* and "boar" for *capra*) and a brief analysis of Bernard's chapter see William Lennox, "Bernard de Gordon on Epilepsy," *Annals of Meaical History*, 3 (1941), 372-383.

[123] See pp. 129-136 above.

[124] "Dicitur eciam quod si scribantur [versus] et portantur ad collum, perfecte curatur. Dicitur eciam quod si pater et mater seu amici ieiunent tribus diebus et postea vadant ad ecclesiam et audiant missam et postea sacerdos fidelis dicat sibi supra capud tercium euangelium quod dicitur in ieiuniis quattuor temporum in vindemiis post festum sancte Crucis ubi dicitur, Spumens et stridens et hoc genus demonum eicitur cum ieiunio et oracione etc. [Mark 9: 17-28] et postea quod ille sacerdos postquam de nocte per intencionem legerit supra capud quod scribat et quod portetur ad collum curatur perfecte procul dubio": *Lilium medicine*, fol. 33v.

examination of two features that are not immediately obvious, namely his implicit reservation and his sense of despair. First, a note of skepticism may be detected when he consistently attached the phrase "*ut dicitur*" or "according to hearsay" to the magical cures, rather than recommending them as tried by his own experience.[125] Actually, frustration and despair in his own experience seemed to account for his listing of the *empirica*. He confessed humbly that

> I have had under my care many patients, young and old, rich and poor, men and women, with almost every kind of epilepsy, and still I have never seen any one cured either by myself or by someone else unless [the patient] was a child or [the disease] had resulted from a bad regimen. ... Yet I was diligent in every detail, and the patients were most cooperative. I do not know the reason, only God knows.[126]

The alleged effectiveness of the cures was, he therefore admitted, contradicted by his personal experience. This point he avowedly emphasized so that the physician would not make "vain and false promises to cure epilepsy. For almost every case of epilepsy is eradicated only with great difficulty, granted indeed that it can be cured at all."[127] Bernard's preference for rational treatment and his reluctant adoption as a last resort of empirical and, a fortiori, of magical cures may also be inferred from his statement about certain virulent skin diseases: "*because* they are cured by divine rather than by human hand, *therefore* experimental remedies have to be used."[128]

If Bernard allowed for magical cures when he found all the natural treatments of epilepsy to be frustratingly ineffective, he did not rely on supernatural causes for its explanation. His etiology was strictly natural

[125] The Gospel cure was also described by Gualterus Agulinus (fl. 1250), who claimed "et nos probavimus illud," but cited Constantinus Africanus (*Pantegni* 5.17) as his source: *Summa medicinalis* 11, ed. Paul Diepgen (Leipzig, 1911), p. 109. Most of Bernard's cures, including some omitted here, are also found in Gilbertus Anglicus' *Compendium medicine* (Lyon, 1510), fol. 109r.

[126] "Tamen dico vobis de epilensia quod ego habui in cura mea multos iuvenes et senes, divites et pauperes, viros et mulieres, et fere de omni specie et tamen nec per me nec per alium vidi aliquem curatum nisi esset puer aut quod provenisset ex malo regimine ... et tamen fui diligentissimus in omnibus et pacientes obedientissimi. Ego nescio causam, Deus scit": *Lilium medicine*, fol. 33r.

[127] "Hoc dico propter hoc quia cum pacientes veniunt ad vos nolite vos ipsos dehonestare cum vanis et falsis promissionibus in curacione epilencie quia fere omnis epilensia cum magna difficultate eradicatur dato quod curari possit": ibid.

[128] "Quia ista ... magis curantur manu divina quam humana, ideo experimentis utendum est": ibid., fol. 10v.

and firmly in keeping with the Hippocratic legacy.[129] Whereas the popular tradition explained the "Sacred Disease" by demonic possession, Bernard attributed it to a blockage of the brain.[130] He thereby exemplified the "remarkable feature of medieval medicine to have kept alive the explanation of epilepsy as a natural disease caused by natural factors."[131] He further discredited several folk tales that were intended to explain mysterious diseases in a supernatural way. Thus, to the popular belief that a patient in a coma (*congelacio*) is "in a trance and is speaking with God and his angels," Bernard replied pointedly that "the kingdom of heaven suffers violence, and the violent take it by force."[132] Usually, his answers were less enigmatic, and he substituted simple natural causes for the supernatural ones of popular explanations. This is seen in the case of wolfish hunger, which the "vulgar" attributed to stepping on an herb that had the magical property of increasing one's appetite; but the true cause, according to Bernard, was simply "inanition" or bodily exhaustion.[133] In another instance, he dismissed the popular belief that depressive mania was caused by a demon. He cited Avicenna's remark that he who teaches medicine (*physicam*) "ought not to concern himself with this" supernatural explanation, since the immediate cause of mania is always natural, namely melancholy or "black bile."[134]

On another medical mystery, namely the nightmare, Bernard critically weighed against each other three lines of traditional explanation. First, "the opinion of the *theologians*," he held, was that nightmares are caused by the demon called incubus. On the other hand, "the *vulgar* say that it is some old woman treading and squeezing the body," and this belief to

[129] The Hippocratic treatise *On the Sacred Disease* opens, "It is thus with regard to the disease called Sacred: it appears to me to be nowise more divine nor more sacred than other diseases, but has a natural cause from which it originates like other affections," tr. Adams, in *The Theory and Practice*, ed. Kelly, p. 347.

[130] *Lilium medicine*, fol. 32v.

[131] Walter Pagel, review of Owsei Temkin, *The Falling Sickness* (Baltimore, 1945), in *Isis*, 36 (1945), 277.

[132] "Vulgares quando vident talem, credunt eum esse raptum, et quod loquatur cum Deo et angelis suis. Sed tamen celum vim patitur et violenti rapiunt illud [Matthew, 11: 12]": *Lilium medicine*, fol. 28r-v.

[133] "Bolismus accidit frequenter itinerantibus et transeuntibus per loca nivosa et frigida, ideo vulgares dicunt quod aliquando herbam calcavit que habet aspectum ad augmentandum appetitum; tamen ista est veritas quod tales sunt cum inanicione corporis": ibid., fol. 63r.

[134] "Dicit Avicenna quod aliquando dicitur quod ista passio [sc. melancholia] provenit a demonio. Et dicit Avicenna quod cum ipse doceat fisicam, de hoc non debet se intromittere, quoniam quomodocumque vadat negocium, causa immediata semper est melancholica, a quocumque fiat transmutacio": ibid., fol. 30r.

Bernard was sheer nonsense: "*nichil est.*" The etiology of the *physicians*, he argued, was the best: "*medici melius opinantur.*" In line with that etiology he then suggested sudden cold as a possible external cause, and corrupted vapors in the brain as an internal cause: both natural explanations.[135] Thus, by his knowledge – the primary quality required by Bernard – the physician might consider himself superior to the layman and to the "vulgar" who, in the Hippocratic phrase, were "ignorant of the true cause."[136] By his diligence and practice according to the art, too, he would distinguish himself from the "common people" who, in another Hippocratic characterization, "are most incompetent, of themselves, to form a judgment how [acute] diseases should be treated."[137]

Although Bernard occasionally invoked common usage as a basis for rejecting the opinion of an author,[138] popular customs and practical opinions were more often the target than the criterion of his criticism. For example, he condemned the practice of "those who prevent the growth of the pubic hair in girls with narcotics and cautery," because that hair was needed to drain vapors.[139] More urgent was his warning for those who "have their uvula incised in order to sing better, because beyond doubt they will never afterwards give a beautiful performance with their vocal instrument."[140] He disapproved, moreover, of such common practices as heavy breakfasts and daytime naps by a group of sedentary persons who could perhaps not be considered "common," namely "wealthy burghers, prelates, scholars (*studiosi*), and the like."[141] The common people, on the other hand, seemed to be ignorant about medicinal baths, since they

[135] "Incubus est nomen demonis, et ideo volunt aliqui quod quando ille incubus directe est supra corpus humanum, et potissime quando iacet et dormit resupinus racione corrupte influencie, aggravans corpus, ita quod videtur pacienti quod suffocetur ... et est opinio theologorum. Vulgares tamen dicunt quod est aliqua vetula calcans et comprimens corpus et hoc [nihil est. Medici autem melius opinantur. Incubus autem provenit] ex causa extrinseca et ex causa intrinseca...": ibid., fol. 32v. The bracketed passage is lacking in London BL MS Harl. 3698, obviously because of a scribal error since it does occur in several other manuscripts I have checked as well as in all the printed editions.

[136] *On Ancient Medicine*, tr. Adams, in *The Theory and Practice*, ed. Kelly, p. 15.

[137] *On Regimen in Acute Diseases*, in *The Theory and Practice*, p. 60.

[138] See, for example, chapter 4, p. 121 above, n. 89.

[139] "Non approbo sentenciam illorum qui impediunt in puellis generacionem pilorum in pudibundis cum narcoticis aut cauterio aut quocumque alio modo, quia illi vapores retenti multa possunt facere vel inducere nocumenta": *Lilium medicine*, fol. 21r.

[140] "Caveant igitur qui in sanitate faciunt incidi sibi uvulam ut melius cantent, quoniam sine dubio cum instrumentis vocalibus nunquam postea faciunt pulchram operacionem. Vigilate igitur": ibid., fol. 49r.

[141] "... sicut faciunt communiter divites in villis et prelati et studiosi et consimiles": *De conservacione*, 4: *Regimen sanitatis*, fol. 66r. Also, on daytime sleeping: ibid., fol. 67v.

administered such baths for paralysis before washing the body. In this, Bernard asserted, "the vulgar are mistaken, and thus the disease becomes worse."[142]

If popular practices often resulted in harm to the patient, also "popular opinion (*opinio vulgaris*)" could severely handicap the physician's intervention. Isaac Judeus, in his *Guide for Physicians* (early tenth century), had bristled at the "foolish opinions" which he found "widespread among the people."[143] In the same vein, Bernard often expressed his impatience with the poor judgment of the common people. He lashed out at the "importunity of the bystanders" who, believing that his therapy was "malicious and cruel," did not allow him to treat lethargic patients. He claimed that he had "frequently accused the bystanders of homicide,"[144] even though he admitted that his drastic "shock therapy" was merciless and violent.[145] He was also alarmed to see how "the vulgar erroneously believe" that coition is beneficial and even a cure for leprosy. In addition he warned that he saw not only that "people err shockingly in the phlebotomy of lepers" but also, "too often, that physicians aim at nothing less than drawing all the blood."[146]

Bernard further turned his attention to another popular fallacy which, he suggested, might interfere with the practice of the conscientious physician. This was the belief that patients should not be bled or phlebotomized at certain times, either on particular calendar dates, or on

[142] "Vulgares tamen decipiuntur in isto balneo quia faciunt ante corporis mundificacionem, et ideo morbus augmentatur": *Lilium medicine*, fol. 35v.

[143] *Guide for Physicians*, tr. Saul Jarcho, in *Bulletin of the History of Medicine*, 15 (1944), 187.

[144] "Multi [litargici] moriuntur propter assistentes, credentes quod hec sit impietas et malicia, et non permittunt talia fieri ... et frequenter accusavi assistentes de homicidio quia non permittebant me operari ... et multi eciam moriebantur et potissime illi in quibus non poteramus operari propter importunitatem assistencium": *Lilium medicine*, fol. 27r.

[145] "Et si [paciens litargicus] esset multum profundatus in sompno gravi, sicut ego vidi multos, tunc trahantur pili barbe et pectinis, et trahantur nares et ?fringantur digiti cum duplici strictura vehementi ... sine omni misericordia, et habeatur sues cum porcellis coram facie eius ... et tube sint ibi et campane magne et timpana et ferra magna que cotidie percutiantur cum martello et vasa enea terribiles sonitus faciencia et clamores fortissimos, aliter dormiendo moriuntur": ibid., fol. 27r. See also n. 98 above. A parallel, slightly different treatment was prescribed by Bartholomeus Anglicus who acknowledged, "all this I have drawn from Platearius and Constantine": *De proprietatibus rerum* 7.7, tr. Walsh, p. 483.

[146] "In lepra coitus non competit, licet sit vulgaris opinio et erronea, quod non solum confert sed curat ... dico vobis quod turpiter video homines errare in flebotomia leprosorum, multociens vidi quod ad nihil aliud intendebant medici nisi quod totum sanguinem trahunt; vigilate ergo": *Lilium medicine*, fols 15v-16r.

the so-called Egyptian days, or during an unfavorable lunar phase. Bernard refuted those who "claim to have some secret pamphlets (*libellos apocrifos*)" proscribing venesection on certain calendar days, and he scoffed that "to observe such things is pure ignorance, sensationalism (*curiositas*), and malice."[147] Another belief, which he labelled as "error and sensationalism," held that the physician should not phlebotomize on Egyptian days, that is on the presumed anniversaries of the days on which God had sent the plagues upon Egypt.[148] In an elaborate refutation Bernard argued that the plagues were neither natural nor universal. If any "malign influence" had been exerted by the stars, it was due to a "supernatural" intervention and it ceased once the punishment had accomplished its purpose, "as the teachers of Sacred Scripture determine." Moreover, the punishment affected only the Egyptian subjects of the Pharaohs but not the Jews in Egypt, much less people of other regions and eras.[149] Therefore, Bernard concluded, these things are recorded in the calendar and preserved by the Church only for commemoration, and "we should not observe any of them either with regard to phlebotomy or with regard to any treatment of the human body."[150] He conceded, however, that the phlebotomy might "have to be postponed" if the barber or "some other meddler (*curiosus*)" mentioned the Egyptian day to the patient because then, "if anything adverse should happen, it would all be blamed

[147] "Aliqui enim dicunt se habere aliquos libellos apocrifos qui tractant de flebotomia secundum istum modum ... ista autem nullo modo sunt observanda quoniam aspectus orbis benivolus aut malivolus non habet ita generalem et fixam et determinatam influenciam ... ideo talia observare est ignorancia pura et curiositas et malicia quare ista quantum nunc ego possum apprehendere nullo modo sunt observanda": *De conservacione*, 1: *De flebotomia*, fol. 22r-v.

[148] "Aliqui dicunt quod dies egyptiacas debemus observare in flebotomia facienda pro tanto quod in kalendario scribuntur, et dicunt quod adhuc est aspectus malivolus illa hora.... Error et curiositas consistit in observacione illarum dierum": ibid., fol. 22r. For a brief synopsis of the history of Egyptian days, see Thorndike, *Magic*, 1: 685-688.

[149] "Dies egyptiacas penitus non debemus observare quoniam illa plaga id est percussio non fuit naturalis nec ex aliquo aspectu orbis sed fuit supernaturalis et eciam non fuit universalis sed quantum ad populum pharaonis qui morabatur in egypto, quod patet quia illa percussio non cadebat nec super iudeos ibidem existentes nec super illos qui erant in grecia et ita de aliis regionibus ... cum illa plaga fuerit supernaturalis finita illa operacione rediit aspectus ille malevolus ad naturam precedentem. Si aliquid materiale corrumpens fuit ibi sicut stella regia, suo cursu peracto rediit in materiam preiacentem sicut volunt doctores sacre scripture," *De conservacione*, 1: *De flebotomia*, fol. 22r.

[150] "Dico quantum occurrit materia de presenti quod nichil de illis debemus observare circa flebotomiam nec circa aliquam operacionem applicandam corpori humano ... nulla observacio est facienda, ex hijs que scribuntur in kalendario nullus deberet moveri, quoniam ecclesia servat et scribit ad memoriam tantum ubi propter aspectum malivolum gracianus eciam in decreto prohibet non observari dies egypciacos": ibid.

on the physician"; thus, "although the day itself has absolutely no influence," it may have to be observed "on account of the poor judgment of the vulgar."[151]

With regard to the connection between phlebotomy and lunar phases, Bernard subscribed to the belief that venesection should be avoided at full moon since this was a time of high humidity during which, according to Galen and the experts, putrefaction sets in sooner.[152] His adherence was less categorical, however, when it came to another belief in lunar influence, namely the view that venesection would fail when the moon is in Gemini. Here he gave the impression that his many years of experience gradually made him more skeptical towards popular applications of astrology. Thus, in his earlier *Liber pronosticorum* (1295), he declared that "no one should perform phlebotomy when the moon rises in Gemini."[153] In his treatise *De flebotomia*, composed thirteen years later (1308), he repeated the same rule, on which "the wise astronomers are agreed." By then, however, Bernard's own personal experience had successfully defied that rule:

> It happened to me that I had noted the hour when the moon would be exactly in Gemini. Then I wanted to administer a bloodletting to myself that very hour, but I was not mindful [about the moon]. When everything was prepared, I remembered, but I did not wish to postpone it. I performed the phlebotomy, and it never went better for me.[154]

Bernard also gave another indication of his growing skepticism towards some aspects of astrology when he discussed the preparation of

[151] "Si barbitonsor dicat pacienti quod est dies egyptiaca aut aliquis alter curiosus dicat, tunc oporteret prolongare quia si aliquid accideret totum medico imputaretur. Ubi tamen est necessitas tunc nihilominus facienda esset nec aliquo modo observare debemus nisi propter malum iudicium vulgarium ... licet penitus dies nullam habeat efficaciam": ibid.

[152] "Dicit Galenus 3 de creticis diebus quod corpora mortuorum cicius putrefiunt et dicunt experti quod arbor incisa in plenilunio non est bene durativa ... et viella collecta id est lignum eius in plenilunio numquam erit bene sonora et panni abluti putrefiunt": ibid., fol. 22v. Compare pp. 134-135 and notes 153-154 above on the presumed hemorrhagic effects of moonlight.

[153] "Nullus debet facere flebotomiam luna existente in geminis, quoniam aut nichil exiet de sanguine aut bis percutietur": *Liber pronosticorum*, fol. 101r.

[154] "Conveniunt eciam sapienciores astronomi quod flebotomia non est facienda luna existente in geminis quoniam aut nullus exiet [sanguis] aut bis percutietur aut morietur. Mihi tamen accidit quod ego notaveram horam in qua luna precise esset in geminis; postea volui facere flebotomiam mihi ipsi in illa hora sed non eram memor et tunc quando fuit totum paratum et ego fui recordatus et nolui dimittere et feci flebotomiam et nunquam melius fuit mihi." *De conservacione*, 1: *De flebotomia*, fol. 22v.

compound medicines. In his *Liber pronosticorum* he had advised that "the physician should make all his preparations under definite [astrological] influences."[155] He qualified this advice with an afterthought in the *Lilium medicine*, written some eight years later (1303). After first suggesting that a remedy for gray hair "might perhaps be better if prepared under a favorable planet," he dismissed this suggestion "because it does not directly concern the physician."[156]

In spite of all his expressed doubts and skepticism towards practical applications of astrology in medicine, Bernard still appeared somewhat hesitant in discrediting popular beliefs. This hesitation becomes evident in the chapter of his treatise *De flebotomia* where, at the hand of the personal experience cited above, he challenged the belief in times unfavorable for bloodletting: he concluded the same chapter with the admission, "on something like this I dare not make absolute assertions."[157] In this regard, the late Professor Thorndike remarked that "Bernard defies astrological medicine successfully in personal practice but he continues to preach it in his writings."[158] While this observation reflects the paradox in Bernard's attitude, it should perhaps be qualified in the light of a distinction, not so much between his practice and teaching, but rather between two aspects of medical astrology, namely general theories and practical applications. Bernard adhered to the astrological paradigms in so far as they belonged to the learned tradition, but he was skeptical towards their applications in popular practice and especially towards their misunderstanding by the "vulgar."

It seems surprising that Bernard, who emphasized the physician's superiority over the "vulgar" in knowledge, diligence and professionalism, did not discuss pragmatic aspects of medical deontology, including fees and bedside manners.[159] He thereby differed from such contempora-

[155] "Medicus debet facere omnes confecciones suas in certis influenciis, verbi gracia volo facere medicamen constrictivum frigidum et siccum sicut est diacodyon, certe confeccio debet fieri in simili influencia et complexione, et quia conveniet radix inferior et superior fortissima erit influencia": *Liber pronosticorum*, fol. 101r.

[156] "Forte si esset talis quod esset sub planeta benivolo facta melius esset; sed istud dimitto quia illud directe non competit medico": *Lilium medicine*, fol. 22r.

[157] "De aliquo autem sicut in isto capitulo non audeo perfecte asserere cum nondum meminerim auctoritate alicuius magistri in medicina et ideo istud lectoris industrie derelinquo": *De conservacione*, 1: *De flebotomia*, fol. 22v. See also the reservations suggested in n. 147 above, "quantum nunc ego possum apprehendere," and in n. 150, "quantum occurrit materia de presenti."

[158] Thorndike, *Magis*, 2: 857.

[159] Unless, of course, Bernard was the author of the treatise *De modo medendi* attributed to him and discussed briefly in chapter 3 above, pp. 93-94.

ries as Arnold of Villanova and Henri de Mondeville.[160] He seemed, indeed, less interested in the mundane details of "medical etiquette" than in the basic principles of general ethics,[161] when he declared that "the physician should be a great moralist." In reply to those philosophers who claimed that the study of ethics was their province, Bernard affirmed that it pertained to the physician as well, not only because man's morals affect his health,[162] but also because the physician is a man of the world.

> Because the greatest wisdom is to know how to move among people, and of all the artisans and learned men the physician moves more commonly among the people, therefore he is held to know some moral science.[163]

Thus Bernard applied to the physician's concern the dual rationale of moral knowledge that went back to Plato and Aristotle, the harmony of social intercourse and the well-being of the entire individual.

It is, perhaps, only normal that Bernard found in psychosomatics, that is the interaction between mind and body, a justification for his concern with ethics, or that he often applied psychosomatics in explaining and treating diseases.[164] What deserves to be emphasized, however, is the fact

[160] See Henry E. Sigerist, "Bedside Manners in the Middle Ages; The Treatise *De cautelis medicorum*, attributed to Arnold of Villanova," *Quarterly Bulletin* of Northwestern University Medical School, 10 (1946), 136-143; also Mary C. Welborn, "The Long Tradition: A Study in Fourteenth-Century Medical Deontology," *Mediaeval and Historiographical Essays in Honor of James Westfall Thompson* (Chicago, 1938), pp. 344-357.

[161] By his emphasis on ethics rather than on etiquette Bernard would seem to belong more to the "old school," whereas Arnold and Henri exemplified "the new secular emphasis," outlined by Loren C. MacKinney, in "Medical Ethics and Etiquette in the Early Middle Ages: The Persistence of Hippocratic Ideals," *Bulletin of the History of Medicine*, 26 (1952), 25. I am preparing a more detailed examination of Bernard's place in the context of medical deontology.

[162] "Queritur utrum medicus in quantum medicus teneatur considerare [moralia] et videtur quod non.... Intelligendum ergo quod de istis moralibus possumus loqui dupliciter, aut habendo solum aspectum ad animam, et ita tractat philosophus; medicus autem ulterius in relacione ad corpus ... medicus considerat sanitatem et per consequens omnia que attribuuntur sanitati, et quia sanitas conservatur et amittitur et reducitur per moralia ideo medico pertinet in quantum medicus est ... ista moralia pertractare...." Bernard then gives several examples, cited from Avicenna and Haly, illustrating psychosomatics, and concludes, "ita mala complexio sequitur malos mores ... et ita saltem in relacione ad corpus magnus moralis medicus debet esse": *De conservacione*, 4: *Regimen sanitatis*, fol. 60r.

[163] "Quia summa philosophia est scire inter homines conversari et inter omnes artifices et homines scientes communiter magis conversatur medicus inter homines, ideo tenetur scire aliquam scienciam moralem": ibid.

[164] See, for example, on memory, *Lilium medicine* 2.12; on lovesickness (*amor hereos*), ibid. 2.20; and on melancholia, ibid. 2.19, especially fol. 30v, where Bernard gives, from personal experience, an interesting suggestion for psychosomatic treatment: "vidi enim

that, though apparently a deeply pious man, he often based his ethical thinking on psychosomatics and natural reasons rather than on authority or pious considerations.[165] Thus, as we have seen earlier, he gave no ascetic but only medical arguments against drunkenness, although he concluded that it was "entirely against God and reason."[166] Another example, perhaps somewhat trivial but nonetheless indicative of Bernard's natural premises, is found in his explanation of monastic silence. He thought, namely, that monks did not talk at table in order to concentrate better on their food.[167] It may be added that this emphasis on psychosomatics and natural explanations logically resulted in a Stoic viewpoint in Bernard's ethics. Thus, he stated that, "because the body follows the mind in its actions just as the mind follows the body in its sufferings (*passionibus*)," man should be moderate in everything, avoiding anger and fear, sadness and envy; he should also rejoice moderately, unless he was weight conscious because according to Galen and Avicenna "moderate joy is fattening."[168] At the same time, it is worth noting that Bernard did not mention either love or piety as the sovereign virtue, except for the young student;[169] instead he expressed the view that justice, which in

---

melancolicum et a fortuna cum gladio fuerat vulneratus et ... quam diu fuerit vulnus apertum optime fuit curatus et statim cum vulnus fuit curatum et clausum reversa fuit mania, et puto quod si vulnus fuisset diu apertum et fuisset remediatus sicut oportet, quod curatus fuisset."

[165] The combination of piety and psychosomatics is illustrated in Bernard's description of vertigo: "Ad bonitatem ergo iudicii multum facit temperamentum complexionis et humorum.... Quare eligendus est magis doctus, bone complexionis et habitudinis, ut medicus operans et iudex et socius et omnis quo indigemus, quoniam si corpus temperatum corrumpitur, quid erit de distemperato? Non est igitur aliquid ut videtur quod ita faciat ad bonitatem speculacionis divinorum et humanorum sicut bonitas regiminis et morum": ibid., fol. 27r.

[166] See above, pp. 120-121 and n. 88. Also, on coition: "cui data est licencia secundum legem debet spergere semen in matricem tantum et nullomodo in alio foramine secundum legem et naturam": *Lilium medicine*, fol. 88v.

[167] "Non est bonum cum aliquis loquitur quod comedat quia cibus transiret ad pectus nec e converso quia aer transit ad stomachum, et hoc potest esse una racio quare religiosi non loquuntur in mensa, ne una operacio impediat aliam": ibid., fol. 46r.

[168] "Inter illa que mirifice conservant sanitatem anime et corporis est quod homo in omnibus actibus suis sciat temperate se in accidentibus anime [habere] ... quod etiam non irascitur nec timeat nec tristetur nec doleat nec lugeat nec alicui invideat ... sed cum summa paciencia sustineat.... Quoniam corpus sequitur animam in suis accionibus sicut anima corpus in suis passionibus.... Gaudium temperatum impinguat secundum Galenum et Avicennam": *De conservacione*, 4: *Regimen sanitatis*, fols. 68v-69r.

[169] "Prima et principalis [condicio] quod [discipulus] Dominum Deum diligat et timeat...": ibid., fol. 59r. I hope to publish an annotated edition of Bernard's instructions for adolescent behavior in conjunction with a sequel to my article on "The Idea of

Aristotelian thought regulated the universe, also governed all the other virtues.[170]

From the fact that Bernard's ethics were to a large extent based on nature, inspired by pagan authors, and solidly based on the Hippocratic-Galenic natural etiology and therapeutics,[171] we should not infer that he was a "naturalist" in the theological sense. In fact, he explicitly stressed Saint Augustine's stand against the Pelagian heresy, in which the "ancient philosophers had wrongly believed" that man can "attain beatitude by his natural powers (*ex principiis naturalibus*)."[172] Hence he interwove pagan and Christian viewpoints, and he illuminated them with quotations from Aristotle and the Bible, from Seneca and Saint Augustine, and from Suetonius and Saint Jerome.[173] Because of this combination of classical and theological authorities,[174] as well as because of the harmony between his solid faith in Nature and his pious belief in things beyond Nature, Bernard may justly be considered a "humanist" in his own right. He perhaps even deserves that designation in the Ciceronian sense, not only because of his stoicism, his reliance on nature, and his familiarity with letters, but also because of his interest in the whole human being, both mind and body, and above all because of his deep concern with rational knowledge and its diligent application to practice as foundations of the physician's *officium*.

Childhood and Child Care in Medical Writings of the Middle Ages," *Journal of Psychohistory*, 4 (1977), 461-490.

[170] "Justitia enim legalis est omnis virtus, quae praecipit omnibus aliis; praecipit enim temperantiae ... fortitudini ... prudentiae...": *De tyriaca*, ed. Pagel, p. 222.

[171] For Hippocratic ethics and nature, see Pedro Lain Entralgo, "La etica hipocratica," in *Melemata: Festschrift für Werner Leibbrand* (Mannheim, 1967), pp. 57-63. For Galenic ethics and nature, see L. Meunier, "Essai sur Galien et le Galénisme," *Janus*, 9 (1904), 270-284 and 313-324; and Walther Riese, "La pensée morale de Galien," *Revue philosophique de la France et de l'Étranger*, 60 (1963), 331-346.

[172] "Hanc autem beatitudinem nullus potest adquirere ex suis principiis naturalibus sicut crediderunt et male philosophantes primi...": *De conservacione*, 4: *Regimen sanitatis* prologue, fols. 52v-53r. And, "non quod in hanc felicitatem possint ex principiis naturalibus devenire sicut crediderunt philosophi antiqui sed indigent gracia...": ibid., fol. 53v.

[173] Bernard's combination of pagan and Christian viewpoints is especially striking in the prologue to his book *De conservacione*, 4: *Regimen sanitatis*, of which I hope to publish an analysis and text edition in the near future.

[174] For more on Bernard's classical, literary and theological references see chapter 1 above.

Conclusion

It would be difficult to find a more worthy representative of Montpellier's claim to fame, of the academic tradition, and of the medical profession around 1300 than Bernard de Gordon. He was a devoted though not uncritical member of the community and master of the medical *studium*. He was also an erudite author who, while perhaps not a genius, aimed at making his works as useful as possible. Therefore he was concerned with conciseness and clarity, he organized his subject matter logically, and he covered and where necessary supplemented the medical curriculum. Only occasionally drifting into speculative dialectic, he usually oriented his teaching towards practical applications, and he was appropriately remembered by the subsequent generation more as a practitioner than as a theoretician. In spite of certain shortcomings, such as his deference to bookish authority or his limited reliance on clinical observation, and in spite of some assumptions that may now seem naive, Bernard was, in the context of his own time, a man of science. He used his sources selectively and critically, he applied both reason and experience, he preferred rational explanations and procedures, and he consistently emphasized natural causes. It was not a vain boast when he called himself a "sedulous inquirer," especially since he also acknowledged the limits of his knowledge. By conceiving of medicine as an applied *scientia* and as a rational *ars*, he observed the vital balance between theory and practice and avoided the extremes of academic rationalism and mystical empiricism that have so often been associated with the Middle Ages. By insisting that the physician should be not only knowing and experienced but also diligent and ethical, Bernard lived up to the Hippocratic legacy of medical deontology. In fine, Bernard de Gordon has a lasting appeal as a medieval representative of an age-old tradition of scholarship, a dedicated teacher, an active member of a demanding profession, and especially as a general practitioner distinguished by his humane concerns and learning.

Appendix

Works Composed by or Attributed to Bernard in Manuscripts and Printed Editions

The following list of Gordoniana is appended not only in order to illustrate the renown of Bernard but also with the intention to share these findings with other interested historians and to solicit further references. The list is the incidental result of textual study rather than the product of systematic bibliographical and paleographical research and, as a preliminary tally sheet, it is neither final nor complete. Thus, some libraries that probably possess Gordoniana (especially in Spain) have not yet been investigated; and references to several items listed have not yet been checked in library catalogues.

The appendix combines the manuscripts and printed editions listed in the bibliographical studies on Bernard by von Töply, Sudhoff, Bell, O'Neill, Herndon, and Winau, with those found in reference works such as the *Catalogue of Incipits* by Thorndike and Kibre and the *Dictionnaire biographique* of Wickersheimer, or those mentioned in sundry articles, and also those communicated by thoughtful colleagues. In addition to the authentic works of Bernard and those attributed to him in manuscripts, catalogues, or bibliographies, some items have been included that are inferentially related to either of these categories.

The works are arranged alphabetically according to the first important word of their title, thereby ignoring the most common terms such as *Capitulum*, *Liber*, or *Tractatus*. The titles are numbered in order to facilitate cross references. The title of each work is followed by the *incipit*, if known. Listed for each work are, first the manuscripts, alphabetically according to the place (city and library) where they are found, and chronologically for holdings of the same library. Next follow the printed editions, listed chronologically. Last come the translations, arranged alphabetically according to language. Each item has the date and foliation in so far as known and (in parentheses) the source of reference, except for a few instances in which I myself have accidentally discovered the item.

Paleographical data, such as provenance or measurements, have been omitted since these would have made the appendix inordinately long.

Symbols and Abbreviations

Complete references to articles and books will be found in the bibliography.

*	manuscript or edition consulted for this study
Bell	H. I. Bell, "The Bibliography of Bernard de Gordon's 'De conservatione vitae humanae'," *xvIIth Internat. Congress of Medicine* (London, 1913).
des.	*desinit*: last words of the text.
expl.	*explicit*: note added to text by author or scribe (equivalent to a colophon in printed editions).
GW	*Gesamtkatalog der Wiegendrucke*, Band 4 (Leipzig, 1930).
Hansen	communicated by Professor Bert Hansen, University of Toronto.
Herndon	Thomas Herndon, "Bernard de Gordon Manuscripts in the United States," *Bulletin of the History of Medicine*, 34 (1960).
HL	*Histoire littéraire de la France*.
Inc.	*Incipit*.
Kibre	communicated by Professor Pearl Kibre, The City University of New York.
MacKinney	Loren C. MacKinney, "Extant Medical Miniatures in Treatises in Manuscripts prior to c. 1550," in his *Medical Illustrations in Medieval Manuscripts*.
McVaugh	communicated by Professor Michael R. McVaugh, The University of North Carolina at Chapel Hill.
O'Neill	Ynez Violé O'Neill, "Dates in the Printed Editions of the *Lilium Medicinae*," *SA*, 49 (1965), 86-89; and "The History of the Publication of Bernard de Gordon's *Liber de Conservatione Vitae Humanae*," *SA*, 49 (1965), 269-279.
NYAM	at the New York Academy of Medicine.
Pansier 1	Pierre Pansier, "Les maîtres de la Faculté de médecine de Montpellier au Moyen Âge," *Janus*, 9 (1904).
Pansier 2	——, "Catalogue des MSS médicaux de France," *SA*, 1 (1908), and
Pansier 3	——, idem, *SA*, 2 (1909).
Sarton	George Sarton, *Introduction to the History of Science*.
SA	*Sudhoffs Archiv* or *Archiv für Geschichte der Medizin*.
s.d.	no date given.
Steinschneider	Moritz Steinschneider, *Die hebräischen Übersetzungen des Mittelalters*.

Sudhoff 1	Karl Sudhoff, "Zur Schriftstellerei Bernhards von Gordon und deren zeitlicher Folge," *SA*, 10 (1917).
Sudhoff 2	——, "Das dem Magister Johannes de Conflente gewidmete Regimen Acutorum des Bernhard von Gordon," *Mitteilungen zur Geschichte der Medizin*, 14 (1915).
Sudhoff 3	——, "Mittelalterliche Einzeltexte zur Beulenpest," in *Historische Studien zur Natur- und Heilwissenschaften* (Berlin, 1930).
Thorndike	Lynn Thorndike, *A History of Magic and Experimental Science*.
TK	Lynn Thorndike and Pearl Kibre, *Catalogue of Incipits*, and
TKa	Ibid., Addenda and Corrigenda.
Wickersheimer	Ernest Wickerheimer, *Dictionnaire biographique des médecins en France au Moyen Âge*.
Winau	Rolf Winau, "Ein Lilium medicinae-Handschrift aus der Besitz des Nikolaus von Kues," *Medizinhistorisches Journal*, 1 (1966).

1. *Affectus preter naturam curandi methodus*
   Inc. "Ingenia curandi morbos sunt decem..."
   Manuscript:
   Paris Arsenal 708, 14th c., fols. 232-234 (Pansier 1: 10).
   Identical with Bernard's *De ingenio sanitatis*.
2. *Antidotarius* (*Liber*)
   Inc. "Digestivum. In passionibus frigidis capitis recipe..."
   Manuscript:
   Paris *BN 6966, a. 1461, fols. 1r-3r (Wickersheimer 1: 75).
   Identical with Bernard's *Lilium medicine* 7.21-24.
3. *Apostematibus* (*Capitula compilata de*)
   Inc. "Curacio apostematum in generali. In primis apponuntur..."
   Manuscript:
   Kassel Q.med.8, ca. 1418, fols. 75v-87v (McVaugh).
   27 chapters drawn primarily from Gilbertus Anglicus' *Compendium* and Bernard's *Lilium medicine* 1.19-22 and 3.2.
4. *Aquis* (*De*)
   Inc. "Capitulum de curatione cuiuslibet tabis [rather than *talis*, as in TK 188]..."
   Manuscript:
   Oxford *Corpus Christi 125, 15th c., fols. 14r-15v (see also Thorndike 2: 205, n. 2).
   This is chapter 6 of Bernard's *De marasmode*.

5. *Arbor egritudinum*
 Inc. (prol.) "Grata laboranti labor omnis bravia confert..." (text) "Egritudo morbus seu infirmitas est mala qualitas..."
 Manuscript:
 Vatican *Pal. 1083, 23 April 1458, fols. 308v-319r (TK 494). *Expl.* "Scriptus et finitus per Johannem [de Büdingen] in preclaro studio Heidelbergensi."

Archanum, see *Translacio*

6. *Astronomia medicorum*
 Manuscript:
 Cambridge Univ. 1705 (Wickersheimer 1: 75).
 Probably identical with Bernard's *Liber pronosticorum* 5.7.

7. *Authenticis urinarum* [*signis*] (*Opusculum de*)
 Inc. "Urina quinquies distinguitur, scilicet ratione..." (TK 1609).
 Manuscripts:
 Cambridge Trinity 500, 15th c., fols. 1v-3r.
 Oxford *Bodl. can. misc. 455, 15th c., fols. 239r-242va (not -249 as in TK).
 Although attributed to Bernard in the MSS, this diagnostic compendium of uroscopy is radically different in form, content and use of sources from Bernard's *De conservacione*, 2: *De urinis*, as well as from the treatises in items 9 and 10 below.

8. *Cautela preservandi a contagio leprosorum* (*De*)
 Inc. "Signa lepre ex Avicenna Gordonio Gerardo Viatico..." (TK 1502).
 Manuscript:
 Vatican *Pal. 1207, 15th c., fols. 15r-16v.

9. *Cautelis necessariis medico ad urinas* (*Nota de*)
 Inc. "Primo pete semper cuius est. Si dicit mulieris..." (TK 1109).
 Manuscript:
 Oxford *Bodl. can. misc. 455, 15th c., fols. 237r-238va (incomplete).
 Much of the text is identical with passages in the tract on how to avoid deception in uroscopy, *De cautelis medicorum*, ascribed to Arnold of Villanova and discussed by Henry E. Sigerist in the *Quarterly Bulletin* of Northwestern University Medical School, 20 (1946), 136-143.

10. *Cautelis urinarum* (*De*)
 Inc. "Prima: Urina hominis cognoscitur quatuor modis..."
 Manuscript:
 London Wellcome 548, 15th c., fol. 48r-v (Moorat 412).
 Inc. "Cum urina tibi ad iudicandum fuerit presentata..." (TK 350).
 Printed editions:
 with Bernard's *Lilium medicine*; *Venice, 1498 (NYAM); Paris, 1542 (NYAM).
 with Bernard's *De urinis*, Ferrara, 1487 (*GW* 4086).

11. *Cibis* (*De*)
 Inc. "Cum omnis cibus detur per comparacionem ad virtutem et morbum et

accidencia, ideo oportet aliquid de istis breviter pertractare. Morborum quidam sunt peracuti... / ... [*des*.] et istud posset esse occasio et quasi quedam tabula ad faciendum unum tractatum de cibis."

Manuscript:

Paris *BN 16189, a. 1314, fols. 191v-192r (Wickersheimer 1: 75).

12. *Codicillus seu de morbis et eorum curacionibus*

Inc. "Circa omnium egritudinum genera triplex modus..." (TK 217).

Manuscript:

Oxford *St. John's 197, 14th-15th c., fols. 127r-254v (Wickersheimer 1: 75).

This head-to-toe compendium ends with fevers, whereas ordinarily fevers precede particular diseases in Bernard's arrangement. The treatise may be identical to the *Liber medicine*, with the same *incipit*, attributed to Petrus Musandinus.

13. *Compendium compositionum*

Inc. "Quoniam de melioribus amicis quos habere videor..." (TK 1270).

Manuscript:

Cambridge *Gonville and Caius 373 (593), 14th c., fols. 101r-102v.

Bernard's authorship would seem confirmed by the fact that, in this early MS, the work occurs between two authentic Gordonian tracts, the *Regimen acutorum* and *De gradibus*. However, it also occurs between two works by Bernard but is called "practicella ex dictis Mesue abbreviata a magistro Johanne de Parma" in Paris MS *BN 16189, a. 1313-1314, fol. 188r. Works with similar incipits are also attributed to several other authors (see TK 1270, 1295, 1244, and TKa 101).

14. *Compendium regiminis acutorum* (see also *Regimen acutorum*)

Inc. (prol.) "Cum (Quoniam tamen) omnis prolixitas sit noverca veritatis ideo ego intendens (videns) librum regimenti diffuse late quoque conditum, fui motus amore mei socii magistri Johannis de Confluente hunc librum a capite breviter sub quodam epilogo pertractare."

(text) "Regimen acutarum consistit in tribus..."

Manuscripts:

Basel *D.I.11, ca. 1400, fols. 128v-129v (TK 1304).

Erfurt *Amplon. F.236, a. 1361, fols. 222r-223r (Sudhoff 3: 93).

—— *Amplon. F.237, mid-14th c., fol. 85r.

Paris *BN 16189, a. 1313-1314, fols. 187v-188r.

Vatican *Pal. 1083, a. 1458, fols. 281v-283r.

—— *Pal. 1229, 15th c., fols. 251r-252v (TK 215).

Printed editions:

with Bernard's *Lilium medicine*, *Venice, 1498 (NYAM); Paris, 1542 (NYAM); Frankfurt, 1617 (NYAM).

Translation:

Hebrew: by anonymous translator (four MSS listed without references, Steinschneider 786).

15. *Compillacion*
Inc. "Quant aucun veut faire aucune bonne operation et vouloir [*sic*] compiller et faire euvre salutaire... / ... [*des.*] ilz puissent acquerir le royaume de Paradis."
Manuscript:
Paris *BNfr 1327, 15th c., fols. 1-38 (*HL* 25: 324; Pansier 1: 541; Pansier 3: 388).
Compilation, "faite par maistre B. de Gourdon, docteur en medecine et par lui compillé en la noble université de Montpellier l'an mil trois cens ou mois de juing," of extracts from the *Lilium medicine* and from the *Liber pronosticorum*.

16. *Confectione ducente ad insenescibilitatem* (*Capitulum de*)
Inc. "Colligantur in primis quinque aque..."
Manuscript:
Erfurt *Amplon. Q.227, early 14th c., fol. 209r-v.
Printed edition:
*Sudhoff 1: 186-187.
The text coincides with the last chapter (7) of Bernard's *De marasmode*.

17. *Conferentibus et nocentibus* (*De*)
Inc. "Conferunt cerebro fetida (ut) in gravi..." (TK 246).
Manuscripts:
Cambridge Peterhouse 178, 13th c., fols. 202v-207v.
Oxford *Bodl. can. misc. 455, 15th c., fols. 226r-237r.
Also attributed to Galen, Johannes Platearius, Gualterus Agulinus, Arnold of Villanova, and so on.

18. *Conferentibus et novicis* (*De*)
Inc. "Nocet cerebro argentum..." (not "cerebrum argento," as in TK 914).
Manuscript:
Oxford Oriel 4, 14th c., fols. 226-(229).

19. *Conservacione vite humane* (*De*)
This work comprises four parts, 1: "De flebotomia," 2: "De urinis," 3: "De pulsibus," and 4: "Regimen sanitatis." When the parts are treated as separate treatises, as is often the case, they will be found under their respective entries. Listed here are only the MSS and printed editions that contain all four or at least three of the parts.
Inc. (prol.) "Homo (enim) est princeps et rex omnium animalium..."
"Flebotomia est evacuacio utilis (universalis)..."
Manuscripts:
Birmingham, University of Alabama Reynolds Library 5007, mid-14th c., fols. 1-24 (Herndon 81).
Cambridge *Gonville and Caius 373 (593), 15th c., fols. [46r-87r] (TK 563, 756, 884, 1420; the text is in disarray in this MS, which is of generally poor quality and contains parts 1, 2 and 4).

Cues *308, 14th c., fols. 27-65v.

Erfurt *Amplon. F.237, mid-14th c., fols. 88r-121v (Sudhoff 1: 172-174).

—— *Amplon. Q.227, mid-14th c., fols. 230r-299v (Sudhoff 1: 172, 186).

—— *Amplon. F.236, 14th c., fols. 67r-93v (Sudhoff 1: 172, 181; this MS contains parts 1, 2 and 3. On fol. 115v, after Bernard's *Liber pronosticorum*, Amplonius Ratinck von Berka (1363–1435) has added, "Nota preter istos libros adhuc quosdam alios composuit, viz. Lilium et librum de conservatione sanitatis [i.e. 'Regimen sanitatis'], quos alibi habeo.")

—— *Amplon. F.267, mid-15th c., fols. 28r-96v.

Krakow Jag. 2110 (parts 1, 2 and 4; McVaugh).

Leipzig 1183, ca. 1400, fols. 100-115 and 147-179 (parts 1, 2 and 4; Sudhoff 1: 172; McVaugh).

London BL Sloane add. 26841, 14th c., fols. 1-71v (Bell 330).

—— BL Sloane 334, 14th c., fols. 218-290 (parts 1, 2 and 4; Bell 333).

—— BL Sloane 512, 14th c., fols. 270-374 (parts 1, 2 and 4; Bell 332).

—— *BL Sloane 3096, 14th c., fols. 324-330, 339-344 (parts 1, 2 and 4; Bell 333).

—— BL Sloane 217, a. 1428, fols. 8-105 (Bell 330).

—— BL Sloane 2481, a. 1470, fols. 2-86v (Bell 330).

Munich CLM 3073, 15th c., fols. 50r-179v (TK 637, 1420).

Paris Arsenal 708, 14th c., fols. 107-155v (Pansier 1: 541 and 2: 10; TK 433, 1150 and 1608).

—— *BN 16189, a. 1313-1314, fols. 121r-170r (O'Neill 274).

Vatican *Pal. 1174, 14th c., fols. 9v-73v (TK 637).

—— *Pal. 1083, a. 1447-1458, fols. 199v-273r (TK 1150, 1420 and 1608; in this MS the parts and chapters are in total disarray: compare item 32 below).

Printed editions:

complete, in *Opus Lilium medicine*, Lyon: Guillelmus Rovillius, 1574, pp. 669-855 ("the most complete printed compilation of Bernard's writings": O'Neill 278).

Parts 2, 3 and 4, with *Lilium medicine*, *Frankfurt, 1617 (O'Neill 278; NYAM).

Note: several MSS and printed editions contain only part 4, "Regimen sanitatis" (see item 65 below), under the title of the entire work, *De conservacione vite humane* (see O'Neill).

20. *Consilia medica*

Manuscript:

Lawrence, University of Kansas B22, 14th c. (Hansen).

Drawn from Bernard's *Lilium medicine* 1.13-18.

Crisi et criticis diebus (De), see *Liber pronosticorum*

21. *Dosi medicinarum (De)*
Inc. "Cum quidam de melioribus amicis quem haberem rogavit me ut ei in scriptis redigerem breviter..."
Manuscript:
Erfurt *Amplon. F.236, a. 1361, fols. 94r-95v (Wickersheimer 1: 75).
Contrary to the assertion by Wickersheimer, Sudhoff (1: 181-182) did not attribute this work to Bernard, but suggested that it was an excerpt from Mesue Jr.'s *De medicinis laxativis* or a tract by Johannes de Parma (fl. 1350), and that the attribution to "Gordonius" was added in the MS ca. 1500. It may be noted, however, that in this MS the treatise occurs between two authentic Gordon works.

22. *Electuarium preciosissimum*
Inc. "Licet sint multa que ducunt hominem ad ultimum senectutis... / ... [*expl.*] Hec sunt verba domini Bernardi de Gordonio in libello de marasmoude [*sic*]."
Manuscript:
Vatican *Vat. lat. 5377, 15th c., fol. 146r-v.
From chapter 6 of Bernard's *De marasmode*. This extract is listed by Thorndike (*Isis*, 13 (1929), 91) as "Capitulum de curatione cuiuslibet talis [*sic*, tabis]"; compare items 4 and 16 above.

23. *Epidemia (De)* (see also *Regimen tempore pestilentiae*)
Inc. "Epidemia habet esse formaliter a corpore supercelesti..."
Manuscript:
Erfurt *Amplon. Q.219, 13th-14th c., fols. 202r-204v (Sudhoff 3: 45-48).
This text appears different from the following.
Inc. "Homo est rex et princeps..."
Manuscript:
Metz 282, 15th c. (Sudhoff 1: 184).
This is probably an extract from Bernard's *De conservacione vite*: compare the *incipit* of the prologue (item 19 above).

Excerpta ex Galeni et Avicenne libris, see *De ingenio sanitatis*

24. *Experimenta*
Manuscript:
Wolfenbüttel 912, 15th c. (Sudhoff 1: 185).

25. *Extracta*
Translation:
Hebrew: by Ashen, son of Moyses of the Olobrega family, in Arles a. 1468 (Pansier 2: 387).
Manuscript:
Lyon Palais des Arts, Hébreu 15, 15th c., item 6 (Pansier 3: 402).

Febre pestilentiali (De), see *Regimen tempore pestilentiae*

26. *Febribus (De)*
Inc. "Febris est calor naturalis mutatus in igneum..."

Manuscript:

Amiens 603, 14th c., fols. 1-31 (Pansier 2: 10; Sudhoff 1: 184).

Printed edition:

in *De febribus opus sane aureum* (Venice, 1576), fols. 207v-217r (Wickersheimer 1: 75).

This text may be related to Part 1 of Bernard's *Lilium medicine*, but there are several works with similar *incipits* (see TK 551-552).

27. *Flebotomia* (*De*): identical with Part 1 of *De conservacione vite*

Inc. "Flebotomia est evacuacio utilis (universalis) multitudinem humorum evacuans..."

Manuscripts:

Kassel F.med. 5, first half 14th c., fols. 1r-6v (McVaugh).

Naples VIII.D.38, 13th c. [*sic*], fol. 58: illumination of the "Author, Bernardus de Gordonio with book (*De phlebotomia*)" (MacKinney 155).

Oxford Bodl. can. misc. 455, 15th c., fols. 196r-207v (TK 635).

—— St. John's 172, 15th c., fols. 264r-270 (TK 563): *inc.* "Flebotomia est multum iuvativum in tempore sanitatis..."; text begins with chapter 2.

Paris BN 16196, 14th c., fols. 27v-46r (TKa 88).

—— *BN 6971, 15th c., fols. 143r-172v (TK 637).

—— *BN 15117, 15th c., fols. 84v-89r: *inc.* "Dicit Galenus xj de ingenio..."; text begins in the middle of chapter 4.

Vatican Vat. lat. 1331, s.d., fols. 216v-217v (McVaugh).

Translation:

Hebrew: by anonymous translator (from the vernacular), in Castejon d'Ampurias, December 1378 (Steinschneider 787).

28. *Flebotomia* (*De*)

Translation:

Hebrew: translator, place and date unknown.

Manuscript:

Boston Medical Library 16, a. 1468. The text follows the Hebrew translation of Bernard's *Lilium medicine* (Ballard, *Catalogue* 12), and is "different from Part 1 of *De conservacione vite*" (Herndon 81).

Flebotomia (*Tractatus de*), see *Flores Lilij*

29. *Fleur de Chirurgie*

Louis Dulieu, "Une exposition d'ouvrages anciens d'anatomie," *Monspeliensis Hippocrates*, 21 (1966), 29 has an illustration with the caption, "gravure tirée de la *Fleur de Chirurgie* de Bernard Gordon, Paris, Jehan Petit, 1504." I have not yet been able to find this work.

Fleur de lys en medecine, see *Practice dicta Lilium medicine*

30. *Flores antidotarii*
Inc. "Electuaria calida: Dyamargariton..." (compare TK 495, 412).
Manuscript:
Munich CLM 26908, 14th c., fols. 92- (Wickersheimer 1: 75).
Contrary to the assertion by Wickersheimer, Sudhoff (1: 183) did not attribute this work to Bernard but indicated that it was similar to the first part of Johannes de Sancto Paulo, "Cogitanti michi..." (compare TK 229-230).

31. *Floribus dietarum (De)*
Inc. "Corpus hominis ex quatuor constat humoribus..." (Sudhoff 1: 181).
Manuscripts:
Erfurt Amplon. F.275, a. 1300-1310, fols. 91r-94r.
—— Amplon. F.289, early 14th c., fols. 32v-33v.
—— *Amplon. F.237, 14th c., fols. 73v-74v.
—— Amplon. Q.185, s.d., fols. 46r-49v.
Printed edition:
as **Bernardi Gordonii de Floribus dietarum Liber unus*, with the *Lilium medicine*, Lyon 1574, pp. 948-963 (NYAM).
This work was also attributed to Arnold of Villanova (TK 268), but it was most probably composed by Johannes de Sancto Paulo (TK 269; Wickersheimer 2: 480).

32. *Flores Lilij*
"Incipiunt flores lilij bernardi de gordonio compilati ab eo in Monte Pessulano anno 1307 xxij die februarij. [*Inc*.] Flebotomia secundum Avicennam canone primo est evacuacio universalis..."
Manuscript:
Vatican *Pal. 1083, a. 1447, fols. 217r-223r (TK 564).
This text is extracted, not from the *Lilium medicine* as the title might suggest, but from the *De conservacione vite*, 1: *De flebotomia*.

33. *Fluxibus (De)*
Manuscript:
Vatican Pal. 719 (Wickersheimer 1: 75).
This text may be related to Bernard's *Lilium medicine* 5.16, "De recapitulacione omnium fluxuum."

Geomancia, see *Translacio*

34. *Glosse super Viaticum Constantini*
Inc. "Cum omnia ex quatuor elementis... / ... [*des*.] Expliciunt notule magistri Bernardi super Viaticum Constantini."
Manuscript:
Warsaw BOZ-59, 15th c., fols. 302r-431v (Kibre).

35. *Gradibus (De)*
Inc. "Doctrinam de gradibus intendimus triplicem ordinare..."
Manuscripts:
Berlin F.186, 15th c., fols. 84v-89v (Sudhoff 1: 179).

Cambridge *Gonville and Caius 373 (593), 14th c., fols. 103-104v (TK 460).
—— Univ. 1693 (Ii.I.i), 14th c., fols. 1-12 (TK 460).
Copenhagen Bibl. regia Gl.kgl.S.229F, 15th c., fols. 58r-70.
Cues *308, 14th c., fols. 108v-116 (TK 460).
Erfurt *Amplon. F.236, early 14th c., fols. 50v-52r (Sudhoff 1: 179).
—— *Amplon. F.237, mid-14th c., fols. 86-87v (Sudhoff 1: 179).
—— *Amplon. Q.227, 14th c., fols. 304-308r (Sudhoff 1: 179, 186).
Kassel Q.med. 16, 3rd quarter 14th c., fols. 47v-54v (McVaugh).
Leipzig 1183, end 14th c., fols. 180r-182v (Sudhoff 1: 180).
—— 1170 (Sudhoff 1: 180).
London *BL Sloane 3096, 14th c., fols. 330v-335r.
—— BL Sloane 217, a. 1428 (Bell 330).
Oxford Bodl. can. misc. 411, 14th-15th c., fols. 180v-187 (TK 460).
—— *Bodl. can. misc. 455, 15th c., fols. 188-192 (TK 460).
—— Merton 225, 14th c., fols. 90v-102r (TK 460).
Paris Arsenal 708, 14th c., fols. 150-153 (Pansier 2: 10).
—— *BN 16189, a. 1314, fols. 195v-199v.
Vatican Pal. 1115, a. 1430, fols. 168r-175v (TK 460).
—— *Pal. 1083, a. 1447-1458, fols. 273v-278v.

Printed edition:
by Julius Pagel, *"Über die Graden der Arzneien nach einer bisher ungedruckten Schrift des Bernhard von Gordon aus dem Jahre 1303," *Pharmaceutische Post*, 28 (1895), 65-67, 131-133, 142-144, 180-182, 221-225, 257-262.

Translation:
Hebrew: translator, place and date unknown.
Manuscript:
Oxford Bodleian Uri 420 (Steinschneider 787).

36. *Hernia (De)*
Inc. "Herniarum plures sunt species..."
Manuscript:
London *BL Harley 4087, 15th c., fols. 23r-26r (TK 612).
Identical with Bernard's *Lilium medicine* 7.7.

37. *Imaginibus signorum (De)* (see also *Tractatus ad faciendum Sigilla*)
Inc. "Aries est signum femineum [*sic*] igneum..."
Manuscript:
Wiesbaden *79, a. 1518-1520, fols. 54r-56r (TK 131, 723).

38. *Impregnacione (De)* (see also *De sterilitate*)
Inc. "Quia sterilitas (multis) causis (accidit), ideo ab ipsis inchoemus. Temperamentum complexionis est causa fecunditatis et lapsus est causa sterilitatis..."
Manuscripts:
Brussels Koninklijke Bibl. II.1413, a. 1411, fols. 148rb-149vb, *des.* "vitet tunc omnem sollicitudinem et tristiciam et offensas, sit in

gaudio et leticia cum exercicio non violento etc. Est finis Deo gracias. Explicit tractatus gordonii de impregnacione, scriptus et finitus in montepessslano per manum petri de outre xxii die mensis decembris a.d. 1411 amen" (McVaugh).

Paris *BN 6957, 15th c., fols. 55v-57v, *des*. "et fiant pessaria cum lana" (compare Paul Diepgen, "Studien ... Arnald von Villanova," *Sudhoffs Archiv*, 6 (1912), 390).

39. *Ingenio sanitatis (De)*, or *De decem ingeniis curandorum morborum*, or *Tabula ingeniorum curandi morbos*

Inc. "Ingenia curacionis (curandi) morborum (morbos) quantum est de presenti sunt decem..."

Manuscripts:

Boston Medical Library 18, 13th-15th c., fols. 68v-73r (Herndon 81).

Cambridge *Gonville and Caius 373 (593), 14th c., fols. 85r-88v [-125v in old foliation] (TK 743).

Cues *308, 14th c., fols. 106r-108v.

Erfurt *Amplon. Q.219, mid-14th c., fols. 197r-200r (Sudhoff 1: 167 and 3: 46; title in this MS, added in a later hand, "Excerpta ex Galeni et Avicenne libris de ingenio sanitatis a Bernardo de Gordonio deprompta").

—— *Amplon. F.236, early 14th c., fol. 49r-v (Sudhoff 1: 166).

—— *Amplon. F.237, mid-14th c., fols. 85r-86r (Sudhoff 1: 167).

—— *Amplon. Q.227, ca. 1350, fols. 205r-208r (Sudhoff 1: 165).

Leipzig 1183, end 14th c., fols. 142r-145r and 194v-196r [twice] (Sudhoff 1: 167; McVaugh).

—— 1161, s.d., fols. 108v-110 (McVaugh).

—— 1213 (Sudhoff 1: 167).

London BL Sloane 334, 15th c., fols. 297v-301r (TK 744).

Munich CLM 340, a. 1304, fols. 259v-265r (Sudhoff 1: 167).

—— CLM 8808, a. 1426, fols. 19r-22r (Sudhoff 1: 167).

Paris *BN 16189, a. 1313-1314, fols. 189v-191v.

—— BNna 3035, a. 1412-1442, fols. 35v-41v (McVaugh).

Vatican *Pal. 1083, a. 1447-1458, fols. 283r-285v (TK 744).

—— Pal. 1229, 15th c., fols. 253r-253v (Kibre; McVaugh).

Warsaw BOZ-59, 15th c., fols. 431v-436r (Kibre).

Printed editions:

with Bernard's *Lilium medicine*, Ferrara, 1486 (*GW* 4081); Lyon, 1491 (*GW* 4082); Venice, 1497 (*GW* 4083); *Venice, 1498 (*GW* 4084; NYAM); *Paris, 1542 (NYAM); *Frankfurt, 1617 (NYAM).

Translations:

Hebrew: translator, place and date unknown (Steinschneider 786).

Spanish: translator, place and date unknown.

Printed edition: with Spanish *Lilio de medicina*, Seville, 1495 (*GW* 4086).

40. *Instructio brevissima pro iter agentibus*

Inc. "Qui longas itinerationes suscipit, ante omnia corpus per purgationem et venesectionem mundificet..."

Printed edition:

with Johannes Strobelberger, **Galliae politicae et medicae descriptio*, 2nd ed. (Jena, 1621), pp. 288-293 (Wickersheimer 1: 75).

This text differs from the corresponding passages in Bernard's *Lilium medicine* 1.31 and *De conservacione vite*, 4: *Regimen sanitatis* 27.

41. *Lapide philosophico* (*De*)

Inc. "Guidonis duodecimo omnis namque ordinata generatio est..."

Manuscripts:

Cambridge Trinity 1120 (III), 15th c., fols. 23r-26v (TK 592).

Oxford Bodl. Ashmole 1420, 16th c., fols. 71-80 (TK 592).

Lilium medicine, see *Practica dicta Lilium medicine*

42. *Marasmode* (*De*)

Inc. (prol.) "Sed ut servemus consuetudinem nostram antequam veniamus ad execucionem, queramus ... utrum seminalia intrent materialiter constitucionem fetus..."

(text) "Marasmodus est corrupcio viventis corporis in siccitate..." (compare TK 848).

Manuscripts:

Cues *308, 14th c., fols. 136v-139v.

London *BL Sloane 3096, 13th-14th c., fols. 335v-336v.

Oxford *Bodl. Digby 56, 14th c., fols. 52r-55v (Thorndike, "Translations of Works of Galen ... by Niccolò da Reggio," *Byzantina Metabyzantina*, 1 (1946), 226).

—— *Corpus Christi 125, 14th c., fols. 13v-15r, incomplete (McVaugh).

Vatican *Pal. 1234, 14th c., fols. 135r-139v (Kibre).

Wolfenbüttel Helmst. 444, 15th c., fols. 127v-130r (Sudhoff 1: 184).

43. *Medicaminibus* (*De*)

Inc. "Digestivum in passionibus superioris capitis. Recipe utriusque salvie..."

Manuscript:

London *BL Sloane 2527, 15th c., fols. 146v-153v (TK 430).

From the *Antidotarius* of Bernard's *Lilium medicine* 7.21-24.

44. *Medicinis conficiendis* (*De*)

Inc. "Cum volueris conficere medicinam prius consideranda sunt...".

Manuscripts:

Cambridge Gonville and Caius 411, 13th c., fol. 120v (TK 352).

Oxford Bodl. Digby 79, 13th c., fols. 58v-82 (TK 352).

—— Oriel 4, 14th c., fol. 216 (TK 352).

45. *Membrorum (Tractatus omnium)*
"Incipit tractatus brevis et utilis Bernardi de Gordonio omnium membrorum..."
Manuscript:
Cambridge Jesus Qq.12, 15th c., fol. 281 (TK 736).
Probably identical with Bernabus, *Materia medicine*.

46. *Modo medendi (De)* "in egritudinibus presertim cum medicus nescit causam" (not "egritudinibus pristinis," as in TK 318)
Inc. "Cum medicus ut plurimum a principio egritudinis..."
Manuscript:
Vatican *Pal. 1083, a. 1447-1458, fol. 322r-v, "Explicit tractatus bernardi de gordonio de modo medendi antequam cognoscit morbum."

47. *Modus medendi* (also *Modus practicandi*)
Inc. "Ego fili carissime tuis acquiescens precibus medendi breves ysagogas tibi tradere non negligo. In primis itaque tibi trado duos gladios unum estivalem alium hyemalem..."
Manuscripts:
London *BL Sloane 134, 15th c., fols. 170r[old 9r]-172r (TK 487).
Oxford Oriel 4, 14th c., fols. 229-230 (TK 487).

48. *Modus medendi extractus de Modo Coffonis*
Inc. [acephalonus] "vel paucitate medicine cui adhibeatur remedium..."
Manuscript:
Munich *CLM 19901, 15th c., fols. 1r-11r (Sudhoff 1: 182).

49. *Morbi contagiosi*
Inc. "Febris acuta, ptisis, scabies, pedicon, sacer ignis..." (TK 550).
This is actually a sentence in Bernard's *Lilium medicine* 1.22 and 3.3, rather than a treatise.

50. *Morbis puerorum (De)*
Inc. "Passiones puerorum adhuc in cunabulis..." (TK 1028)
Manuscript:
Vatican *Vat. lat. 10213, 15th c., fol. 541r, *des.* "Finis tractatus de egritudinibus puerorum secundum bernardum de gordonio."
The text is virtually identical with that of the late 12th-c. anonymous *Practica puerorum*, ed. by Karl Sudhoff, *Erstlinge der pädiatrischen Literatur* (Munich, 1925), pp. XLI-XLII.

51. *Natura et complexione superiorum (De)* "quantum medico"
Inc. "Medicus velit nolit aut subponendo..."
Manuscript:
Paris *BN 16189, a. 1313-1314, fols. 194r-195v (Wickersheimer 1: 75).
The title is identical with that of Bernard's *Liber pronosticorum* 5.7, but the text is that of 5.8-10.

52. *Natura signorum (De)*
Inc. "Medicus autem velit nolit supponendo aut quoquo modo fit..."
Manuscript:
Cambridge Trinity 1122 (O.II.18), 15th c., fols. 159v-160r (TK 863).
Probably identical with Bernard's *Liber pronosticorum* 5.8-10: compare item 51 above.

53. *Neutralitatibus decidencie (Compendium de)*
Inc. "Considerandum est urina utrum sit remissa vel elevata..."
Manuscript:
Wolfenbüttel Aug.F.12.4, a. 1452, fols. 168-171 (Sudhoff 1: 183).
Printed edition:
by Erna M. Kollert, in **Zwei Compendien 'de neutralitatibus decidencie' von Bernhard Gordon und Johann Tornamira* (Leipzig dissertation, 1924), pp. 15-25.

54. *Niños (El tractado de los)* "con el regimiento dela ama"
Inc. "La muger quando se allegare al parto..." (O'Neill 277).
Printed edition:
with Bernard's **Lilio de medicina*, Seville, 1495 (copy at the Hispanic Society of America, New York City; Hansen).
Spanish translation of Bernard's *De conservacione vite*, 4: *Regimen sanitatis* 1-6 (whereas "original unbekannt" in *GW* 4086).

55. *Passionibus capitis usque ad pedes (De)*
Inc. "Recipe utriusque salvie utriusque..." (*sic*, TK 1339).
Manuscript:
Basel *D.III.1, 15th c., fols. 110r-116r.
This text appears related to the *Antidotarius* of Bernard's *Lilium medicine* 7.21-24.

56. *Passions depuis le chief jusques aux piez (Petit tractié qui parle des)*
Inc. "Vecy ung digestif lequel vault es passions froides di chief... / ... [*des.*] item le livre de graduatione."
Manuscript:
Paris BNfr 1288, 15th c., fols. 136-142 (Pansier 2: 388).
A French translation of Bernard's *Lilium medicine* 7.21-24.

57. *Perah ha-refu'ot* ["Bud of Health"]
Manuscript:
Oxford Bodl. Michener 390 (Steinschneider 785).
Hebrew translation by Moyses ben Samuel of Roquemare, *alias* John of Avignon, made in Seville a. 1360, of Bernard's *Lilium medicine* (Wickersheimer 1: 76; Sarton 3: 873).

Phlebotomia, see *Flebotomia*

58. *Practica dicta Lilium medicine*
Inc. (prol.) "Interrogatus a quibusdam (quodam) Socrates quomodo posset optime dicere..."

(text) "Febris est calor innaturalis (naturalis) mutatus in igneum..."

Manuscripts:

Avignon 993, 16th c. (Pansier 2: 10).
—— 996, 16th c. (Pansier 2: 10).
Besançon 466, 16th c. (Pansier 2: 10).
Bethesda NLM 497, a. 1349, fols. 1-151 (Herndon 80).
—— NLM 498, a. 1419, fols. 3-215 (Herndon 80).
—— NLM 507, 15th c., incomplete (Herndon 80).
Cambridge St. John's 180 (G.12), a. 1312, fols. 1r-204v (TK 772).
—— Gonville and Caius 487 (483), 14th c., fol. 10ra- , *inc.* "[A]postema est inflammatio seu tumor preter naturam..." (rather than "Postea est," as in TK 1066; compare the *inc.* of Bernard's *Lilium medicine* 1.18, and TK 114).
Chartres 224, 16th c. (Pansier 2: 10).
Cues *304, 14th c., fols. 1-199.
Dresden 227, 15th c. (Edmund Güntz, "Aussatz ... Bernhardus Gordonius," *Archiv für Dermatologie und Syphilis*, 2 (1870), 63).
Erfurt *Amplon. F.258, before 1310, fols. 1r-122v (Sudhoff 1: 163).
—— *Amplon. Q.219, a. 1335, fols. 1r-194v, lacks 5.1-17 (Sudhoff 1: 163).
Évreux 114, 14th c., illuminated (Pansier 2: 10; MacKinney 124).
Kassel F.med.7, ca. 1435, fols. 2r-223rb (McVaugh).
Krakow Jag. 2108, 15th c. (McVaugh).
Leipzig 1189, before 1350, fols. 1-109v, illuminated (Sudhoff 1: 164, 182; MacKinney 133).
—— 1190, 14th c., fols. 1-135, illuminated (MacKinney 133).
—— 1191, ca. 1400 (Sudhoff 1: 164).
—— 1192, a. 1434, fols. 1-191 (Sudhoff 1: 164; Sudhoff, "Die Leipziger Kindslagenbilder mit deutschen Texte," *SA*, 2 (1909), 423).
London *BL Harley 3698, "5 February 1305," fols. 1-99 (not a. 1304 as in Winau).
—— BL Sloane 3096, 14th c., fols. 220-232 (Bell 333).
—— *BL Royal 12.D.XIII, 14th c., fols. 17-105v (not -106 as in TK 772).
—— BL Sloane 334, 14th c. (Bell 333; O'Neill 87).
—— BL Sloane 512, 14th c. (Bell 333; O'Neill 87).
—— Wellcome 130, a. 1330, fols. 1-145 (Bell 333; O'Neill 87).
Mainz 520, 14th c. (Winau 27).
—— 523, a. 1417 (Winau 28).
—— 194, 15th c. (Winau 28).
Metz 178, 16th c. (Pansier 2: 10).
—— 381, 16th c. (Pansier 2: 10).
Montpellier 451, 15th c. (Pansier 2: 10).
Munich CLM 13019, a. 1313, fols. 1-110v (Sudhoff 1: 164; TK 772).

—— CLM 340, a. 1304 [?], fol. 51r (TKa 80, *inc.* "Alopicia est casus capillorum cum ulceribus...": *Lilium medicine* 2.1; the MS contains the entire *Lilium* according to Sudhoff 1: 164).
—— CLM 457, s.d. (Sudhoff 1: 164).
—— CLM 3522, s.d. (Sudhoff 1: 164).
—— CLM 26908, 14th c., fols. 1-81r (Sudhoff 1: 164, 183).
Paris Arsenal 708, 14th c., fols. 1-124 (Pansier 2: 10).
—— Arsenal 979 (16.B.L.), 14th c., fols. 25-112 (TK 772).
—— *BN 6964, 14 November 1305, fols. 1-81v (TK 772).
—— *BN 16189, 26 October 1313, fols. 3-121r, *expl.* "Fuit completus liber iste per everardum de oudensale anno incarnacionis domini m.ccc.xiij. feria quinta ante festum apostolorum symonis et jude." (Erroneously listed as BN 16190 by Eduard Seidler, *Heilkunde*, p. 316).
—— *BN 6965, 14th c., fols. 1-120v.
—— *BN 7054A, 14th c., fols. 3r-153r.
—— *BN 11227, 14th c., fols. 1-203r, *inc.* "Fuit quidam interrogatus a quodam quomodo posset..." (Bell 333; O'Neill 87).
—— *BN 15117, 15th c., fols. 1-82r (Bell 333; O'Neill 87).
—— BN 7119, s.d., fol. 144, *inc.* "Lepra est morbus...": *Lilium* extract (McVaugh).
—— Faculté de Médecine, 15th c. (Pansier 2: 10).
Philadelphia College of Physicians, 14th c., formerly Osler Library MS 7523 at McGill University (Herndon 80).
Vendôme 238, 14th c. (TK 772).
—— 176, 16th c. (Pansier 2: 10).
—— 338, 16th c. (Pansier 2: 10).
Vienna 2338, a. 1322, fols. 1-99 (TK 771, 772).
—— 2864, 14th c., fols. 2-363 (TK 771).

Printed editions:
Naples: Francesco del Tuppo, 20 May 1480 (*GW* 4080).
Ferrara: Andreas Belfortis [Gallus], 18 May 1486 (*GW* 4081).
Lyon: Antoine Lambillon & Martin Sarrazin, 2 May 1491 (*GW* 4082).
Venice: Johannes et Gregorius de Gregoriis for Benedictus Fontana, 16 January 1497 (*GW* 4083).
*Venice: Bonetus Locatellus for Octavianus Scotus, 22 December 1498 (*GW* 4084; NYAM).
Venice, 1521 (Sarton 3: 875).
*Paris: Dionysius Ianotius, 1542 (NYAM).
Lyon, 1551 (Sarton 3: 875).
Lyon: Guillelmus Rovillius, 1574 (in "the most complete printed compilation of Bernard's writings," O'Neill 269).
Venice, 1594 (Sarton 3: 875).
*Frankfurt: Lucas Jennis, 1617 ("the last printing" according to Sarton 3: 875, but see the Spanish translation below; NYAM).

Translations:

French: see items 15 and 56.

——: by anonymous translator, Rome a. 1377 (*HL* 25: 328; O'Neill 89).

Printed edition:

La practique qui sappelle fleur de lys en medecine, Lyon, 31 August 1495 (*GW* 4085, as by "[Drucker des Bernardus Gordonio]").

German: by anonymous translator, ca. 1350.

Manuscript:

Vienna Lat. [*sic*] 2864, 14th c., fols. 44v-56r (see below, printed ed. by Carlowitz, from whose introduction it is not clear whether this MS contains a translation of the entire *Lilium medicine* or only of 1.22, "De lepra").

Printed edition:

extract, by Constantin H. Carlowitz, **Der Lepraabschnitt aus Bernhard von Gordons 'Lilium medicinae' in mittelalterlicher deutscher Uebersetzung* (Leipzig, 1913), pp. 14-41.

Hebrew: by Moyses ben Samuel, *alias* John of Avignon, Seville a. 1360 (see item 57).

——: by Jekuthiel ben Solomon, *alias* Bonsenhor Solomon, Narbonne a. 1387 (see item 73).

——: by Leo Joseph, ca. 1400 (Wickersheimer 1: 76).

Manuscripts (respective translators not identified):

Boston Medical Library 16, a. 1468 (Herndon 80).

Dallas, Texas, Dr. Lyle Sellers, 15th c. (Herndon 80).

Irish: "by a member of the O'Hickey family in the northern part of Munster," 15th c., "never published" (Sarton 3: 873; gives no MSS or references).

——: "by Cormac Mac Duinnsthleibe, a distinguished scholar and physician of Ulster," 15th c , "never published" (Sarton 3: 873, gives no MSS or references).

Provençal: provided the basis for the Hebrew translations according to Steinschneider (Steinschneider 786, without further references).

Spanish: *Lilio de medicina*, translator(s) and date(s) unknown.

Printed editions:

*Seville: Meynardo Ungut & Stanislao Polono, 18 April 1495 (rather than 1494 as in *HL* 25: 328; *GW* 4086; O'Neill 87; copy at the Hispanic Society of America, New York).

Toledo: Jan de Villaquiran, 1515 (O'Neill 278 n. 3).

Madrid: Antonio Gonçalez de Reyes, 1697, as *Obras de Bernardo de Gordonio insigne maestro y doctor de Medicina* ("perhaps the most recent printing of the *Lilium*": O'Neill 278).

59. *Pronostico de Montepessulano (De)*
Inc. "Numerus et sufficientia dierum creticorum habetur..."
Manuscript:
Prague 243 (I.F.11), s.d., fols. 176r-182v (TK 960).
Probably identical with Bernard's *Liber pronosticorum* 5.6.

60. *Pronosticorum (Liber)*, or *De crisi et criticis diebus*
Inc. (prol.) "Senectus domina oblivionis est. Nunc autem cum sciencia predicandi sit ardua et difficilis..."
(text) "In morbis pronosticare non possumus nisi morbum cognoscamus..."
Manuscripts:
Bristol City Reference Library, a. 1420-1435, fols. 272v-279v (TK 211, as by Guy de Chauliac; see ed. by Singer, below).
Cues *308, 14th c., fols. 117r-136v.
—— *309, 14th c., fols. 2r-21v.
Erfurt Amplon. F.275, ca. 1300-1310, fols. 95r-117r (Sudhoff 1: 169).
—— Amplon. F.289, before 1350, fols. 33v-43r (Sudhoff 1: 183).
—— *Amplon. Q.227, before 1350, fols. 209v-229v (Sudhoff 1: 168).
—— *Amplon. F.237, mid-14th c., fols. 76r-84r (Sudhoff 1: 168; TK 1428).
—— *Amplon. F.236, end 14th c., fols. 96r-115v (Sudhoff 1: 183).
—— *Amplon. F.267, 14th c., fols. 1r-25r (Sudhoff 1: 169; TK 1428).
Krakow Jag. 818 (McVaugh).
—— Jag. 821 (McVaugh).
Leipzig 1183, 14th c., fols. 31r-44v (Sudhoff 1: 169; McVaugh).
London *BL Royal 12.D.xiii, 14th c., fols. 5r-16r, incomplete: 4.4-5.10; fol. 16v, "Expliciunt quatuor libelli de quibus compilator lilij fecit mencionem in fine lilij."
—— BL Sloane 334, 15th c., fols. 189r-213r (TK 1428).
Munich CLM 26908, before 1350, fols. -91 (Sudhoff 1: 183).
Oxford Magdalen 164, 15th c., fols. 138r-151r, text incomplete or rearranged, since it opens with the *inc.* of the *Liber pronosticorum* 4.1, "Intelligendum est quod in cerebro sunt tres cellule..." (TK 756).
Paris *BN 16189, a. 1313-1314, fols. 170v-185r, 194r-195v.
—— BNna 3035, a. 1412-1442, fols. 148v-194 (McVaugh).
Vatican *Pal. 1174, 14th c., fols. 73v-103r.
—— *Pal. 1083, a. 1458, fols. 285v-308r (TK 692).
—— Pal. 1331, 15th c., fols. 198r-216v (TK 1428).
—— Pal. 1098, 15th c., fols. 187v-224r (TK 692).
Warsaw BOZ-59, 15th c., fols. 255r-290r (Kibre).
Wiesbaden *56, 15th c., fols. 62r-83v (TK 1428).
Wolfenbüttel 81.4.Aug.F., s.d., fols. -383v (Sudhoff 1: 169).

Printed editions:

with Bernard's *Lilium medicine*, Ferrara, 1486 (*GW* 4081); Lyon, 1491 (*GW* 4082); Venice, 1497 (*GW* 4083); *Venice, 1498 (*GW* 4084; NYAM); *Paris, 1542 (NYAM); Lyon, 1574 (O'Neill 269); *Frankfurt, 1617 (NYAM).

ed. by Charles Singer, "Sur un manuscrit attribué à Guy de Chauliac," in *Comptes rendus du 2e Congrès international d'histoire de la médecine* (Évreux, 1922), pp. 168-176. From Bristol City Ref. Libr. MS (see above), *inc.* "Circa istam materiam est intelligendum quod natura ... [TK 211] / ... aliter non poterit esse perfectus in sciencia predicandi nec eciam operandi... / ... [*des.*] Ysopus prassium et liquiricia et sic de aliis et sic preservabitur si deus voluerit": a careful comparison with Bernard's *Liber pronosticorum* leaves no doubt that this text is merely a rearrangement of that work, rather than a treatise by Guy de Chauliac as Singer believed.

Translations:

French: see item 15 above.

Hebrew: by Leo Joseph, ca. 1400 (Wickersheimer 1: 76).

——: translator and date unknown (Steinschneider 787, no MSS listed).

——: translator and date unknown, though different from the preceding (Steinschneider 787, no MSS listed).

Manuscript (apparently of the translation by Leo Joseph):

Boston Medical Library 17, 15th c., fols. 1r-18r (Herndon 81).

61. *Pulsibus* (*De*) (identical with Part 3 of *De conservacione vite*)

Inc. "Pulsus est nuncius qui non mentitur..."

Manuscript:

New York, *NYAM, 14th c., single vellum folio, with beautifully illuminated initial P: a physician taking the pulse of a patient.

Printed editions:

with *Lilium medicine*, *Venice, 1498 (*GW* 4084; NYAM).

with Bernard's *De urinis*, Ferrara: Andreas Belfortis [Gallus], 4 March 1487 (*GW* 4086).

62. *Pulsibus* (*Compendium de*)

Inc. "Sciendum quod pulsus grossus est..."

Manuscript:

Vatican Pal. 1098, s.d., fols. 224r-227v (TK 1398).

From the *incipit* it appears doubtful whether this text is related to Bernard's *De conservacione vite*, 3: *De pulsibus*.

63. *Questio disputata*

Inc. "Questio. Queratur utrum seminalia intrent substanciam fetus et videtur quod non..."

Manuscript:
Erfurt *Amplon. Q.227, 14th c., fols. 208r-209r (Sudhoff 1: 185).
This text is identical with the *questio* that introduces and comprises the first half of Bernard's *De marasmode*.

64. *Ratio computandi per digitos*
Inc. "Hic incipit prima computatio, VI nonas Aprilis signat..."
Manuscript:
Cambridge Univ. Mm.IV.41, 14th c., fol. 100r (TK 620).

65. *Raucedine (De)*
Manuscript:
Prague 1610 (Wickersheimer 1: 75).
Perhaps related to Bernard's *Lilium medicine* 4.3, "De raucedine vocis."

Reductio geomancie, see *Translacio*

66. *Regimen acutorum morborum*, or *Regimen acutarum egritudinum* (see also *Compendium regiminis*)
Inc. "Regimen acutarum egritudinum consistit in tribus..."
Manuscripts:
Basel *D.I.11, ca. 1400, fols. 122v-123v.
Cambridge *Gonville and Caius 373 (593), 14th c., fols. 97r-99r (TK 1344).
—— Univ. 1708 (IV.i.16), 14th c., fols. 54r-60v (TK 1344, as anon.).
Chartres *393, 14th c., fols. 1r-4v (TK 1344).
Cues *308, 14th c., fols. 103r-106v (TK 1344).
Erfurt *Amplon. Q.227, ca. 1350, fols. 202r-205r (Sudhoff 3: 46).
—— *Amplon. F.236, early 14th c., fols. 49v-50v.
—— *Amplon. F.236, a. 1361, fols. 220r-222r (TK 1344, as anon.).
—— *Amplon. F.237, mid-14th c., fols. 84r-85r.
—— *Amplon. Q.219, 14th c., fols. 200r-204v (Sudhoff 1: 168 and 3: 46).
Leipzig 1183, end 14th c., fols. 193r-194v (Sudhoff 1: 167).
—— 1161 (Sudhoff 1: 167, McVaugh).
—— 1170 (Sudhoff 1: 167).
—— 1200 (Sudhoff 1: 167).
Munich CLM 4395, 15th c., fols. 206r-210v, *inc.* "Acutarum egritudinum regimen consistit in tribus..." (TK 29).
—— CLM 340, s.d., fols. 252v-259v (TKa 89).
Paris *BN 16189, a. 1313-1314, fols. 185r-187v.
—— *BN 7124, a. 1488-1489, fols. 223r-258r, *inc.* "Nota secundum Gordonium in libro suo regiminis acutorum [not "regimentis acutorum," as in TK 943] ait ut infra. In morbis enim pronosticare non possumus nisi morbum cognoscamus...": rather than from Bernard's *Regimen acutorum*, the quotatio is from his *Liber pronosticorum*. The excerpt, listed as a "commentary" (TK 943), is actually one of many extensive passages from various Bernardian works

found throughout the entire codex, a compilation on fevers under the title *Collectanea medica*, written between 1488 and 1491 in Dijon by "Hectorem artium et medicine doctorem," Hector Demonstaldo (Wickersheimer 1: 272).

Vatican *Pal. 1083, a. 1447-1458, fols. 278v-281v.

Vienna 5306, 15th c., fols. 132r-134v (TK 1344, as anon.).

Warsaw BOZ-59, 15th c., fols. 290r-296v (Kibre).

Printed editions:

with Bernard's *Lilium medicine*, Ferrara, 1486 (*GW* 4081); Lyon, 1491 (*GW* 4082); Venice, 1497 (*GW* 4083); *Venice, 1498 (*GW* 4084; NYAM); Venice, 1521; *Paris, 1542 (NYAM); Lyon, 1574; *Frankfurt, 1617 (NYAM).

Translation:

Spanish: translator, place and date unknown.

Printed edition, with *Lilio de medicina*, Seville, 1495 (*GW* 4086).

67. *Regimen sanitatis* (identical with Part 4 of *De conservacione vite*, whose title was frequently applied to it)

Inc. (prol.) "Secundum Aristotelem (intencionem Aristotelis) primo ethicorum vita humana est triplex..."

(text) "Mulier cum appropinquat ad partum debet ire..."

Manuscripts:

Copenhagen Bibl. regia Gl.kgl.S.229F, 15th c., fols. 1r-57v.

London BL Sloane 3097, 28 Dec. 1311, fols. 1r-61r (O'Neill 273).

—— BL Sloane 3689, 14th c., fols. 1-56 (Bell 334-335).

—— BL Sloane 513, 14th c., fols. 293v-345r (TK 884).

Metz 178, 14th c., fol. 12r, *inc.* "Nullus debet sumere cibum nisi stomachus sit vacuus..." (TK 957): *Regimen* 11.

New York *NYAM, 14th c., single vellum folio, with beautifully illuminated initial S: a teacher with book on lectern, and three tonsured students at his feet.

Oxford Merton 47, 15th c., fols. 1-49 (TK 1420).

Paris *BN 16190, 14th c., fols. 1r-40v (O'Neill 274), colophon, 15th c., "... iste valens dominus et magister qui compilavit presentem librum non nominat in hoc volumine nomen suum. Revera circa sanitatem corporis ipse melius ordinabit quam magister Arnaldus de Villanova"; and, 16th c., "Videtur quod sit Gordonius ut patet in capitulo xij ... ubi ait, '... sicut dictum est in lilio.' Et etiam in folio 24..."

—— *BN 6971, 15th c., fols. 173r-228v.

Vienna 5315, 15th c., fols. 1r-70v (TK 1420).

Printed edition:

by Joachim Baudisius, as **De conservatione vitae humanae*, Leipzig: Johannes Rhamba & Ernestus Vogelin, 1570 (Sudhoff 1:

174 n. 1; O'Neill 277). In his preface Baudisius states that this treatise, which he brought "ex Academia Montis Pessuli" with other Gordonian MSS, "longe est praeferendus" not only above the rest of Bernard's works but even above practical writings of the 16th century. Bernard's reputation may be reflected in the fact that a copy of this edition (in my possession) is bound between treatises by Vesalius' pupil Levinus Lemnius (*De miraculis occultis naturae* and *De vita ... recte instituenda*, Antwerp: Christophorus Plantinus, 1574) and the *Consilium novum de pestilentia* by the Parisian physician Petrus Droetus (Paris: Andreas Wechelus, 1572).

Translations:

Hebrew: translator, place and date unknown.

Manuscript:

Madrid Escorial Pl.3.c.9 (Steinschneider 787).

It is not clear whether this translation comprises only the *Regimen sanitatis* or Bernard's entire work *De conservacione*.

Spanish: see item 54 above.

68. *Regimine sanitatis* (*Compendium de*)

Inc. "(Et) (licet) per Dei graciam de regimine sanitatis longiorem tractatum intendamus edere... Circa regimen sanitatis cadunt septem consideraciones..."

Manuscripts:

Basel D.I.18, 14th c., fols. 104v-107r (TK 518, *inc.* "Et licet perdidi racionem de regimine...").

London *BL Sloane 420, 14th c., fols. 99v-101r-104 (TK 1031, 220).

—— BL Sloane 3468, 14th c., fols. 27r-31r (TK 828).

This text is identical with Bernard's *Lilium medicine* 5.8.

69. *Regimen tempore pestilentiae*, or *De febre pestilentiali* (see also *De epidemia*)

Inc. "Febres pestilentiales sunt ille que adveniunt..."

Manuscripts:

Glasgow 323 (U.7.17), a. 1452, fol. 1v (TK 550, as anon.).

Munich *CLM 4119, 15th c., fols. 443v-453v (TK 550, as anon.).

This collection of diagnostic materials and prescriptions opens with an excerpt from Bernard's *Lilium medicine* 1.10, "De febribus pestilentialibus," but it also contains quotations from Galeatius de Sancta Sophia, Arnold of Villanova, Gentile da Foligno, Gullielmus Placentinus, and Guy de Chauliac.

70. *Reumate* (*Excerpta de tractatu de*)

Inc. "Signa reumatis calidi acuti secundum Avicennam tertio Can..."

Manuscript:

Vienna *2426, 14th c., fols. 143v-144r (Sudhoff 1: 184).

Only the therapeutic section of Bernard's chapter on rheuma in the *Lilium medicine* 3.16 is excerpted by the anonymous author who claims that he has frequently applied ("sepius sum expertus") several of Bernard's cures.

71. *Secret des dames* (*Le*), "Dix règles pour quoi l'on peut entendre pourquoy les femmes ont chascun mois leurs fleurs selon M. Bernard de Gourdon"
Inc. "Cy commence le secret des dames que peut entendre légèrement le gouvernement des fleurs des dames..."
Manuscript:
Paris BNfr 19989, 15th c., fol. 42 (Pansier 3: 388).
Perhaps a translation of excerpts from Bernard's *Lilium medicine* 7.8-9.

72. *Secreta ex Lilio extracta*, or *Vademecum*
Inc. "Alopicia est casus capillorum cum ulceribus..."
Manuscript:
Kassel, Q.med.2, ca. 1450, fols. 3r-117v (McVaugh).
Even though this text opens with the *inc.* of Bernard's *Lilium medicine* 2.1, it probably contains the entire *Lilium* as rearranged at Padua in 1441 by Wermboldus Overstege de Campis (not "Oustege" as in TK 85).

73. *Shoshan ha-refuah* ("Lily of Medicine")
Hebrew translation of Bernard's *Lilium medicine*, by Jekuthiel ben Solomon, *alias* Bonsenhor Solomon, at Narbonne in 1387 (Sarton 3: 873).
Manuscripts: 14 listed, s.d. (Steinschneider 785).

74. *Sigilla et ymagines contra infirmitates diversas* (*Tractatus ad faciendum*)
Inc. "Incipiamus cum Dei adiutorio de hijs que possunt fieri sole exeunte in quolibet signo... Aries est signum masculinum igneum incisum membris...
Manuscript:
Vienna *3162, 15th c., fols. 239r-241r (TK 723).
Compare item 37 above.

75. *Simplas medicinas et compostas* (*La Summa del Gordo de*)
Manuscript:
Paris Arsenal 2525, s.d., item 8, "copié par Lacurne de St Palaye du MS Vatican 4797" (Pansier 3: 401).
This Provençal text is possibly a translation of the *De medicinis conficiendis* (see item 44 above) or from the *Antidotarius* of Bernard's *Lilium medicine* 7.21-24.

76. *Sterilitate mulierum* (*De*) (see also *De impregnacione*)
Inc. "(Et) quia causa (cause) sterilitatis accidit (accidunt), ideo ab ipsa (ipsis) inchoemus. Temperamentum complexionis est causa fecunditatis..."
Manuscripts:
Cambridge Peterhouse 101.III, 13th-14th c., fol. 5 (TK 1556).
Erfurt *Amplon. F.236, a. 1361, fols. 223-224 (TK 522; though not attributed to Bernard in this MS, the text follows his *Compendium regiminis acutorum*).
Leipzig F.1216, ca. 1400, fols. 137r-138v (Sudhoff 1: 184-185).
Metz 282, 15th c. (Sudhoff 1: 184; TK 1556).
Paris *BN 7066, a. 1441, fols. 28r-32r (TK 1556; Pansier 1: 541 erroneously gives BN 7064).

Vatican *Pal. 1083, a. 1458, fols. 320r-321v, *des.* "Et sic est finis tractatus de sterilitate editus a bernardo de gordonio decano preclari studii montispessulani. Finitus et scriptus per Johannem Borner de Büdingen in octava assumpcionis B. Marie Virginis [22 August] anni etc. [*sic*, 14] 58" (see TK 1214, not as by Bernard).

Tabula ingeniorum, see *De ingenio sanitatis*

77. *Termini medicinales*

Inc. "Et primo de allopicia est..."

Manuscript:

Wilhering 71, 15th c., fols. 272-279 (TK 520).

Probably related to Bernard's *Lilium medicine* 2; compare item 72 above.

78. *Theriaca (De)*, or *De tyriaca*

Inc. (prol.) "(Inquit magnus medicus Hamet Averroys,) Tyriaca est confectio magna curans omnem (venenosam) egritudinem..."

(text) "Queritur utrum tyriaca possit exhibere corporibus sanis..."

Manuscripts:

Cues *308, 14th c., fols. 113r-116v (TK 1574).

Erfurt *Amplon. F.270, 13th-14th c., fols. 73v-76r (TK 1574).

—— *Amplon. Q.227, 14th c., fols. 300r-303v (TK 1574).

Krakow Jag. 825 (McVaugh).

London *BL Sloane 3096, 13th-14th c., fols. 337r-338v (TK 1574).

Paris Arsenal 707, 13th-14th c., fols. 89-92, *inc.* "Inquit ... Averroys postquam prius gratias Deo egero dicam quod quidam ex amicis..." (Sudhoff 1: 180; Pansier 2: 10; compare *incipits* of items 13 and 21 above).

Vatican *Pal. 1083, 15th c., fols. 322v-326v (TK 1574).

Note: some other MSS, in which the text is attributed to Averroës (TK 750), may contain, rather, the treatise by Bernard.

Printed editions:

with *Lilium medicine*, Lyon, 1491 (?, Sarton 3: 875, but not in *GW*).

with Magninus Mediolanus, *Regimen sanitatis*, Florence, 1500 (TK 750, not clear whether Averroës' or Bernard's *De theriaca*).

with introduction and German translation, by Julius Pagel, *"Über den Theriak nach einer bisher ungedruckten Schrift des Bernhard von Gordon," *Pharmaceutische Post*, 27 (1894), 205-206, 220-224, 235-237, 257-262, and the Latin text of the last chapter (6) in Julius Pagel, *Die angebliche Chirurgie des Heben Mesue Junior* (Berlin, 1893), p. 144.

Translation:

German, see the preceding edition.

79. *Translacio, Archanum de reductione geomancie ad orbem*

"Incipit archanum magni Dei revelatum Tholomeo regi Arabum [see TK 727].

[*Inc.*] Per hoc presens opus possunt cognosci..."

Manuscripts:

Cambridge Magdalen 27, 14th c., fols. 166-173, incomplete (TK 1032).

Florence Plut. 89.sup., cod. 34, s.d., fols. 11r-20v (TK 1032).

London *BL Harley 2404, 15th c., fols. 49r-59v (TK 1032).

Paris *BN 15353, 13th-14th c., fols. 87r-92v (Kibre).

—— *BN 7349, 15th c., fols. 138v-141r, incomplete (Wickersheimer 1, 75).

—— BN 7457, s.d., fols. 83r-96r (TK 1032)

80. *Urinis* (*De*) (identical with Part 2 of *De conservacione vite*)

Inc. (prol.) "Dilexi veritatis scienciam investigare..."

(text) "Urina est colamentum sanguinis et (similiter) ceterorum humorum ex natura actionibus natum..."

Manuscripts:

Brussels Koninklijke Bibl., II.1413, 15th c., fols. 91-112 (McVaugh)

Cambridge Trinity O.VIII.2, 13th c. [*sic*], fol. 80r-v, inc. "Urina alba vel glauca vel pallida..." (TK 1507): probably chapter 11.

Kassel F.med.5, first half 14th c., fols. 6v-8v, incomplete: chapters 1-5 (McVaugh).

—— Q.med.16, 3rd quarter 4th c., fols. 28r-47v (McVaugh).

Krakow Jag. 821 (McVaugh).

—— Jag. 2027 (McVaugh).

—— Jag. 2425 (McVaugh).

London BL Sloane 59, 15th c., fols. 156-181 (Bell 334).

—— BL Harley 1612, s.d., fol. 14v- , *inc.* "Regiones urine sunt quatuor. Prima regio est..." (TK 1346): probably chapter 4.

Los Angeles University of California Benjamin Collection 3, s.d., fols. lv- (O'Neill 276).

Metz [178?], 13th c. [*sic*] (Pansier 2: 10).

Paris BN 6860, 14th c., fols. 191r-217v (Kibre).

—— BN 6972, s.d., fol. 90r, *inc.* "Color urine quidem est qui signat temperamentum..." (TK 235, as by Arnold of Villanova): perhaps chapter 2.

Vatican Pal. 1265, 15th c., fols. 25r-55r (TK 1608).

Printed editions:

with Bernard's *Lilium medicine*, *Venice, 1498 (*GW* 4084; NYAM).

with Bernard's *De pulsibus*, Ferrara: Andreas Belfortis [Gallus], 4 March 1487 (*GW* 4086).

Translation:

Italian, by Silvano Chelini and Aldo Scapini, *Il 'De urinis' di Bernardi di Gordon (XIII-XIV sec.)*, Pisa, 1968.

81. *Urinis* (*De*)

Inc. "Intelligendum quod color creatur a vite via..."

Manuscript:

Cambridge *Gonville and Caius 373 (593), 14th c., fols. 72v-74v (TK 756).

This text is different from that of Bernard's *De conservacione vite*, 2: *De urinis*.

82. *Urinis (De)* (commentaries "In librum de urinis Bernardi de Gordonio")

Manuscripts:

Oxford Magdalen 164, 15th c., fols. 127-138, *inc*. "Utrum de urina possit esse..." (TK 1643).

—— All Souls 80, 15th c., fol. 22r-v, *inc*. "In libro super urinas magistri Bernardi de Gordonio..." (TK 690).

Vademecum, see *Secreta*

83. *Visitacione infirmorum (De)*

Inc. "Cum omnis prolixitas sit noverca veritatis..."

Manuscript:

Paris *BN 15373, 14th c., fol. 162r-v (TK 327).

Identical with Bernard's *Compendium regiminis acutorum*.

Bibliography

The following annotated bibliography lists the principal works consulted in the preparation of this study, without claims either to exhaustive completeness or to narrow selectiveness. Omitted, for example, are some modern works that deal with the general medical tradition, with the broader intellectual context, or with the background of medieval society, as well as most incunabular editions of those Hippocratic, Galenic, and other texts that constituted the standard literature of a university-trained physician. Included, on the other hand, are several studies that pertain to important aspects of Bernard's scholarship but that were not cited in my footnotes in order to concentrate on the specifically pertinent references.

The usual arrangement of primary sources and secondary works has been subdivided so that materials with direct bearing upon Bernard de Gordon may be more readily identified.

I. PRIMARY SOURCES

A. Bernard de Gordon Works in Manuscripts and Printed Editions.

See the preceding Appendix.

B. Other Primary Sources

Aegidius Corboliensis or Gilles de Corbeil (ca. 1140–1224). *De laudibus et virtutibus compositorum medicaminum*. Ed. by Louis Choulant in *Carmina medica*, pp. 47-198. Leipzig: Teubner, 1826.

——. *De urinis*. In Bruges MS 467, 13th c., fols. 62-64.

——. *Viaticus de signis et sinthomatibus egritudinum*. Ed. by Valentin Rose. Leipzig: Teubner, 1907.

Alphonsus Cordubensis (fl. ca. 1350). *Epistola et Regimen de Pestilentia*. Ed. by Karl Sudhoff in *Archiv für Geschichte der Medizin*, 3 (1909), 222-226.

Arnold of Villanova (ca. 1240–1311). *Aphorismi de gradibus*. Ed. by Michael R. McVaugh in *Arnaldi de Villanova Opera Medica Omnia*, vol. 2. Granada [and] Barcelona: Universidad de Barcelona, 1975.

This careful critical edition has an elaborate and very important introduction on medicinal degrees.

——. *De conferentibus et nocentibus*. In Bruges MS 468, 14th c., fols. 2-3; also in Bruges MS 470, a. 1275, fols. 109v-113r.

——. *De confortatione visus*. Ed. by Pierre Pansier in *Mémoires de l'Académie de Vaucluse*, 3 (1903), 1-19.

——. *De considerationibus operis medicine*. In Paris MS BN lat. 17847, 14th c., fols. 57-81.

I am preparing an annotated edition of this treatise from this and twelve other MSS.

——. *Experimenta*. Ed. by Michael McVaugh (excerpts) in *Journal of Medieval and Renaissance Studies*, 1 (1971), 107-118.

——. *De improbatione maleficiorum*. Ed. by Paul Diepgen in *Archiv für Kulturgeschichte*, 9 (1911), 385-403.

——. *De sterilitate*. Ed. by Karl Sudhoff in *Archiv für Geschichte der Medizin*, 6 (1913), 380-389.

——. *De vinis*. Tr. by Henry E. Sigerist. *The Earliest Printed Book on Wine*. New York: Schuman's, 1943.

——. *Hec sunt Opera Arnaldi de Villa Nova ... nuperrime recognita ac emendata....* Ed. by Thomas Murchius. Lyon: Fradin, 1509.

Articella. Venice: Baptista de Tortis, 1487. (At the New York Academy of Medecine.)

This collection, related to the "ars medicine," contains the standard curricular works of Hippocrates, Galen, and others in the Latin versions that were available to Bernard.

Avicenna. *Canon medicinae*. Venice: Petrus Maufer, 1486. (At the New York Academy of Medicine.)

——. *Canon*, Book 1. Tr. by Oskar Cameron Gruner. *A Treatise on the Canon of Medicine of Avicenna, Incorporating a Translation of the First Book*. London: Luzac, 1930.

——. *Canon*, Part 1. Tr. by Mazhar H. Shah. *The General Principles of Avicenna's Canon of Medicine*. Karachi: Naveed Clinic, 1966.

While convenient for quick consultation, both of the above translations are defective in several ways.

——. *Canticum*. Tr. by Haven C. Krueger. *Avicenna's Poem on Medicine*. Springfield, Ill.: Charles Thomas, 1963.

Bartholomeus Anglicus (fl. 1220–1250). *De proprietatibus rerum*. Nürnberg: A. Koberger, 1483. (At the New York Academy of Medicine.)

——. *De proprietatibus rerum*, Book 7 on medicine. Tr. with notes and an introductory essay by James J. Walsh in *Medical Life*, 40 (1933), 449-602.

Cartulaire de l'Université de Montpellier. Ed. by the Commission du Cartulaire with introd. by Alexandre Germain. 2 vols. Montpellier: Ricard, 1890.

Chirurgia brevis de apostematibus et vulneribus (early 13th c.?). Ed. by Karl Sudhoff in *Studien zur Geschichte der Medizin*, 11-12 (1918), 265-296.

This is an anonymous Montpellier gloss on Roger of Salerno.

Collectio Salernitana; Ossio documenti inediti e trattati di medicina appartenenti alla scuola medica Salernitana. Ed. by Salvatore de Renzi. 5 vols. Naples: Filiatre-Sevezio, 1852-1859.

Constantinus Africanus (ca. 1015–1087). *Pantegni.* Ed. and tr. by Marco T. Malato and Umberto de Martini. Costantino l'Africano: *L'Arte Universale della Medicina (Pantegni) Parte I - Libro i.* Rome: Istituto di Storia della Medicina dell'Università di Roma, 1961.

Galenus, Claudius. *Opera Omnia.* Ed. by Karl G. Kühn. 20 vols. in 22. Leipzig, 1821-1833.

——. *De sanitate tuenda.* Tr. by Robert M. Green. *Galen's Hygiene.* Springfield, Ill.: Charles Thomas, 1951.

——. *De usu partium.* Tr. with introduction and commentary by Margaret T. May. *Galen on the Usefulness of the Parts of the Body.* 2 vols. Ithaca, N.Y.: Cornell University Press, 1968.

Gautier Agilon, see Gualterus Agulinus.

Gerardus de Solo or de Bierne (fl. 1335–1371). *Introductorium iuvenum, sive de regimine corporis humani in morbis.* In Paris MS BN lat. 7062, a. 1428, fols. 1-31.

——. *Practica super nono Almansoris Rasis.* In Paris MS BN lat. 7062, a. 1428, fols. 32-261.

Gilbertus Anglicus (early 13th c.). *Compendium medicine.* In Bruges MS 469, a. 1271, fols. 1-244.

——. *Compendium medicine sive Laurea anglica.* Lyon: Michael de Capella, 1510. (At the New York Academy of Medicine.)

Gilbertus Cancellarius Montispessulanus (ca. 1250?). *Experimenta.* Ed. by Pierre Pansier in *Janus*, 8 (1903), 20-25.

Gilles de Corbeil, see Aegidius Corboliensis.

Girardus Bituricensis or de Bourges (early 13th c.). *Commentum in Viaticum Constantini.* In Bruges MSS 359, 13th-14th c., fols. 246-277; 467, 13th c., fols. 1-49; and 466, 14th c., fols. 1-125r.

Gualterius Agulinus or Gautier Agilon (fl. 1250). *Compendium urinarum.* Ed. by Julius Pfeffer. Berlin: Schade, 1891.

——. *Summa medicinalis.* Ed. by Paul Diepgen. Leipzig: Barth, 1911.

Guillelmus de Biterris or Gausberti (fl. 1300–1320). *Informatio scolaribus suis* or *Isagoge ad practicam medicine.* In Paris MS BN lat. 7061, 15th c., fols. 125-132 (rather than -126r, as in TK).

Guillelmus de Brixia or de Brescia (fl. ca. 1274–ca. 1326). *Questiones de tyriaca.* Ed. by Michael McVaugh in "Theriac at Montpellier, 1285-1325." *Sudhoffs Archiv*, 56 (1972), 113-144.

——. *Chirurgia.* Ed. by Karl Sudhoff in "Eine Chirurgie Wilhelmus von Brescia (Guglielmus Corvi)." *Beiträge zur Geschichte der Chirurgie im Mittelalter*, 2. Teil, Studienheft 12 (1918), 419-421.

Guillelmus de Congenis (fl. ca. 1215). *Chirurgia.* Ed. by Karl Sudhoff in "Domini et magistri Willelhelmi de Congenis, Burgensis in Montpellier 'Scriptum cirurgie', sowie dazugehörige 'Notulae cirurgie' eines Schülers." *Beiträge zur Geschichte der Chirurgie im Mittelalter*, 2. Teil, Studienheft 12 (1918), 297-384.

——. *Cyrurgie*, Middle English. Ed. by Margaret S. Ogden. *The Cyrurgie of Guy de Chauliac*; Vol. 1. *Text*. Publication of the Early English Text Society no. 265. London: Oxford University Press, 1971.

The companion volume, which will contain an introduction, commentary and a glossary, is in preparation and will further enhance the value of this edition as a boon to scholarship.

Henricus de Hermondavilla or Henri de Mondeville (d. 1317). *Chirurgia*. Ed. by Julius Pagel. *Die Chirurgie des Heinrich von Mondeville (Hermondaville) nach Berliner, Erfurter und Pariser Codices*. Berlin: Hirschwald, 1892.

——. *Chirurgia*. Tr. by E. Nicaise. *La Chirurgie de maître Henri de Mondeville; Traduction française avec notes, une introduction, et une biographie*. Paris: Alcan, 1893.

——. *Chirurgie* (French). Ed. by A. Bos. *La Chirurgie de maître Henri de Mondeville; Traduction contemporaine de l'auteur, publiée d'après le* MS. *unique de la Bibliothèque Nationale*. 2 vols. Paris: Didot, 1897-1898.

Hippocrates. [*Works*. Ed. with] tr. by W. H. S. Jones. *Hippocrates*. The Loeb Classical Library nos. 147-150. 4 vols. Cambridge, Mass.: Harvard University Press, 1923.

——. [*Works*. Tr. by Francis Adams.] Ed. with introduction by Emerson C. Kelly. *The Theory and Practice of Medicine*. New York: Citadel, 1964.

A very convenient edition of the major Hippocratic writings in the 1849 translation.

Isaac Judeus (fl. ca. 900). *Musar Harofim* [*Guide for Physicians*]. Tr. by Saul Jarcho in *Bulletin of the History of Medicine*, 15 (1944), 180-188.

Isidore of Seville (ca. 560–636). *The Medical Writings*. Tr. with introduction and commentary by William D. Sharpe. Philadelphia: American Philosophical Society, 1964.

Johannes Anglicus or John of Gaddesden (ca. 1280–1361). *Rosa anglica practica medicinae*. Ed. by Nicholaus Scilliacus. Pavia: Franciscus Girardengus, 1492. (At the New York Academy of Medicine.)

Johannes Hake von Westerholt (fl. 1314–1349). *Recepte ad preservandum ab epydemia* and *Recepte contra epidemeam*. Ed. by Karl Sudhoff in *Archiv für Geschichte der Medizin*, 5 (1912), 36-39.

Johannes de Tornamira or Jean de Tournemire (ca. 1330–ca. 1390). *Breve compendium de neutralitatibus decidencie*. Ed by Erna Margarete Kollert. *Zwei Compendien 'de neutralitatibus decidencie' von Bernard Gordon und Johann Tornamira*. Leipzig dissertation: Leiner, 1924.

——. *Clarificatorium ... in speculatione curationis morborum cum cura brevi et electa eorundem* [or, *super nono Almansoris Rasis*]. In Paris MS BN nouv. acq. lat. 1399, a. 1438, fols. 1-181.

——. *Parva practica* "quam fecit pro quodam bachalario eunti ad practicam." In Paris MS BN lat. 7061, 15th c., fols. 1-5.

Jordanus de Turre or de la Tour (fl. 1313–1335). *De mulierum impregnacione*. In Paris MS BN lat. 7066, a. 1441, fols. 6-11r.

——. *Recepte facte filio suo eunti ad practicam anno 1318*. In Paris MS BN nouv. acq. lat. 1391, fol. 71r-v.

Liber instrumentorum memoralium; Cartulaire des Guillems de Montpellier. 2 vols. Montpellier: Martel, 1884-1886.

Magistri Salernitani nondum editi. Ed. by Pietro Giacosa. Turin, 1901.

Supplements the *Collectio salernitana* ed. by S. de Renzi.

Nicholas de Montepessulano or de Polonia (fl. ca. 1275). *Antipocras*. Ed. by Karl Sudhoff. "Antipocras; Streitschrift für mystische Heilkunde in Versen des Magisters Nikolaus von Polen." *Archiv für Geschichte der Medizin*, 9 (1916), 31-52.

——. *Experimenta*. Ed. by J. W. S. Johnsson. "Les *Experimenta magistri Nicolai*." *Bulletin de la Société française d'Histoire de la Médecine*, 10 (1911), 269-290.

Petrus Aponensis or Peter of Abano (1250–ca. 1316). *Conciliator differentiarum philosophorum et praecipue medicorum*. Venice: Scotus, 1496. (At the New York Academy of Medicine.)

Petrus Borelli de Capitestagno or de Capestang (fl. 1299–1330). *Cura contra dispositionem ad paralisim*. Ed. by Ernest Wickersheimer in *Bulletin de la Société française d'Histoire de la Médecine*, 18 (1924), 103-106.

Petrus Hispanus or Pope John XXI (ca. 1215–1277). *Thesaurus pauperum*. Venice, 1543 (Italian translation). (At the New York Academy of Medicine.)

Petrus de Nadillis or de Nadilz (fl. 1370–1375). *De impregnacione mulierum*. In Paris MS BN lat. 7066, 15th c., fols. 13-20.

Petrus Vitalis or Vidal (fl. 1318). *Novum kalendarium*. In Paris MS BN lat. 7420A, ca. 1332, fols. 101-108.

This kalendar was drawn up at Montpellier in 1318 and dedicated to Pope John XXII (r. 1316–1324).

Profatius Judeus or Jacob ben Makhir (ca. 1236–ca. 1305). *Almanach sive tabule*. In Paris MS BN lat. 7408A, 14th c., fols. 2v-11r.

Drawn up for the longitude of Montpellier, and beginning 1 March 1300 (or 1301).

——. *Almanach*. Ed. by J. Boffito e Melzi d'Eril. Florence, 1908.

——. *De armillis*. In Paris MS BN lat. 7295A, 14th c., fols. 35-40.

Written in Montpellier at the request of Bernard de Gordon.

——. *De aspectibus lune*. In Bruges MS 523, 14th c., fols. 26v-28v; also in Ghent MS 5 (416), 14th c., fols. 152r-154r.

——. *Quadrans novus*. In Bruges MS 523, 13th-14th c., fols. 46r-62r (rather than fols. 53v-(64r) as in TK).

——. *Quadrans novus*. Ed. by J. Boffito e Melzi d'Eril. Florence, 1922.

——. *Safiha* or *Saphea*. Ed. by J. Millas-Vallicrosa. Barcelona, 1933.

This is Profatius' translation, with the assistance of Johannes Brito or Brixiensis, of the treatise of al-Zarqal.

Raymundus Lullus or Ramon Llull (ca. 1235–1315). *De sterilitate*. Ed. by Walter Artelt and Julius Pagel in *Janus*, 8 (1903), 530-537.

The Regimen of Health of the Medical School of Salerno. Tr. by Pascal Parente. New York: Vantage, 1967.

Ricardus Anglicus or de Montepessulano, or Parisiensis or Salerni (fl. ca. 1280). *Anathomia*. Ed. by Karl Sudhoff in *Archiv für Geschichte der Medizin*, 19 (1927), 209-239.

This is the third of the five treatises that comprise the *Micrologus* of Ricardus Anglicus.

Rogerius de Barone (fl. ca. 1260). *Practica*. In Bruges MS 470, 13th c., fols. 41-94r.

This work consists of four treatises, of which the first are known as the *Rogerina maior* and the last is called the *Rogerina minor* or *parva*.

Les Statuts et privilèges des universités françaises depuis leur fondation jusqu'en 1789. Ed. by Marcel Fournier. 4 vols. Paris: Larose et Fourcel, 1890-1894.

Thalamus parvus Montispessulanus. Montpellier: Martel, 1836.

A collection, published from the MSS by an anonymous editor, of medieval documents of Montpellier: customs, oaths, and chronicles.

Valescus de Tharanta (d. 1418). *Practica que Philonium dicitur*. [No place given]: Nicolaus Wolff, 1500. (At the New York Academy of Medicine.)

II. SECONDARY WORKS

A. Works on Bernard de Gordon

Amasuno, Marcelino Valentin. *El 'Lilio de Medicina' de Bernardo de Gordonio. Contribucion al estudio del lenguaje médico español del siglo XV*. Doctoral thesis, announced as to be published by the University of Salamanca in 1973; still only on microfilm in 1978.

I have not yet been able to see this study.

Astruc, Jean. "Bernardus Gordonius, ou plutôt de Gordonio." In *Mémoires pour servir à l'histoire de la Faculté de Médecine de Montpellier*, pp. 176-181. Paris: Cavelier, 1767.

——. "Notice sur Bernard de Gordon." *Éphémérides médicales de Montpellier*, 3 (1826), i-iv.

Rather than by Astruc himself, this note is by the editor of the *Éphémérides*, although based on Astruc's findings.

Bell, H. I. "The Bibliography of Bernard de Gordon's 'De conservatione vitae humanae'." In *17th International Congress of Medicine*, pp. 325-337. London, 1913.

Carlowitz, Constantin Hans. *Der Lepraabschnitt aus Bernhard von Gordons 'Lilium medicinae' in mittelalterlicher deutscher Übersetzung*. Leipzig (dissertation): Peter Nache, 1913.

This is an edition, with introduction and some annotation (but without critical apparatus) of a German translation of the *Lilium*'s chapter on leprosy, made about 1350 and contained in Vienna MS lat. [*sic*] 2864, fols. 44v-56r. On pp. 42-45 of this dissertation the author gives a useful vocabulary of medicines.

Chevalier, Cyr Ulysse. *Répertoire des Sources historiques du moyen âge; Bio-Bibliographie*. Vol. 1. Paris: Picard, 1905, p. 67.

Gives some older bibliographical references on Bernard de Gordon.

Eichholtz, Martin. *Zahnheilkundliches bei Bernhard von Gordon*. Leipzig dissertation, 1923.

Güntz, Edmund. "Aussatz und Syphilis nach den Quellen des Bernhardus Gordonius." *Archiv für Dermatologie und Syphilis*, 2 (1870), 59-77.

The author observes that Bernard's description of leprosy actually gives the symptoms of syphilis. This is for the author an argument for the view that there was syphilis in Europe well before 1492.

Herndon, Thomas. "Bernard Gordon Manuscripts in the United States." *Bulletin of the History of Medicine*, 34 (1960), 80-81.

Kollert, Erna Margarete. *Zwei Compendien 'de neutralitatibus decidenciae' von Bernhard Gordon und Johann Tornamira*. Leipzig dissertation: Leiner, 1924.

Lennox, William Gordon. "Bernard de Gordon on Epilepsy." *Annals of Medical History*, 3 (1941), 372-383.

Littré, Émile. "Bernard de Gordon, Médecin." In *Histoire littéraire de la France*, 25 (1869), 321-337.

McVaugh, Michael. "Nota sobre las relaciones entre dos maestros de Montpellier: Arnau de Vilanova y Bernardo Gordon." *Asclepio*, 25 (1973), 331-336.

O'Neill, Ynez Violé. "Dates in the Printed Editions of the *Lilium Medicinae*." *Sudhoffs Archiv*, 49 (1965), 86-89.

——. "The History of the Publication of Bernard de Gordon's *Liber de Conservatione Vitae Humanae*." *Sudhoffs Archiv*, 49 (1965), 269-279.

Pagel, Julius L. "Über die Grade der Arzneien nach einer bisher ungedruckten Schrift des Bernhard von Gordon aus dem Jahre 1303." *Pharmaceutische Post*, 28 (1895), 65-67, 131-133, 142-144, 180-182, 221-225, 257-262.

Ed., with German tr. and introduction, of Bernard's *De gradibus*, from Erfurt Amploniana MSS Q. [*sic*, rather than F.] 236 and 237, and from Berlin MS Q. [*sic*, rather than F.] 186. I am grateful to Professor Michael McVaugh who loaned me his copies of this and of the following edition, since the journal is most difficult to obtain.

——. "Über den Theriak nach einer bisher ungedruckten Schrift des Bernhard von Gordon." *Pharmaceutische Post*, 27 (1894), 205-206, 220-224, 235-237, 257-262.

Ed., with German tr. and introduction, of Bernard's *De tyriaca*, from Erfurt Amploniana MSS F.270 and Q.227. Contrary to the assertion of Pagel, this treatise may have been printed in 1491 and in 1500 (see Appendix).

——. *Die angebliche Chirurgie des Heben Mesue Junior*. Berlin, 1893.

On p. 144, Pagel gives the Latin text of the last chapter of Bernard's *De tyriaca*, which he omitted in the preceding edition.

Sarton, George. "Lilium Medicinae." In *Mediaeval Studies in Honor of J. D. M. Ford*, pp. 239-256. Cambridge, Mass.: Harvard University Press, 1948.

Briefly examines Bernard's *Lilium*, Gilbertus Anglicus' *Laurea*, and John of Gaddesden's *Rosa*, without elaborating a comparison between these three compendia.

Sudhoff, Karl. "Das dem Magister Johannes de Conflente gewidmete Regimen Acutarum Aegritudinum des Bernhard von Gordon." *Mitteilungen zur Geschichte der Medizin und der Naturwissenschaften*, 14 (1915), 93.

——. "Mittelalterliche Einzeltexte zur Beulenpest vor ihrem pandemischen Auftreten 1347/48." In *Historische Studien und Skizzen zu Natur- und Heilwissenschaften (Festgabe Georg Sticker)*, pp. 39-48. Berlin, 1930.

Concludes, on pp. 45-48, with excerpts from Bernard's *Regimen acutorum*.

——. "Zur Schriftstellerei Bernhards von Gordon und deren zeitlicher Folge; Eine Handschriftenstudie." *Archiv für Geschichte der Medizin*, 10 (1917), 162-188.

Unravels the relative and absolute chronology of Bernard's works on the basis of a comparative study of MSS in German libraries and of incunabula.

Töply, Robert von. "Die chronologische Reihenfolge der Schriften des Bernard Gordon." *Mitteilungen zur Geschichte der Medizin und der Naturwissenschaften*, 6 (1907), 94-95.

Superseded by Sudhoff's study cited above.

Winau, Rolf. "Ein Lilium medicinae-Handschrift aus der Besitz des Nikolaus von Kues." *Medizinhistorisches Journal*, 1 (1966), 24-30.

A most interesting article on the *Lilium* MS (probably not the oldest extant, contrary to Winau's assumption) that once belonged to Nicolaus Cusanus (1401–1464), and that has marginal notes by him: London BL MS Harley 3698, completed on 5 February 1305 (rather than 1304, as Winau states) by a Swabian student at Montpellier, Heinrich Hembat.

B. General Background and Reference

Afnan, S. M. *Avicenna; His Life and Works*. London: Allen & Unwin, 1958.

Aimes, Alexandre. "Un illustre maître montpelliérain Guy de Chauliac, le père de la chirurgie moderne." *Monspeliensis Hippocrates*, 5 (1962), 3-14.

Amadou, Robert. *Raymond Lulle et l'Alchimie; Introduction au Codicille avec notes et glossaire*. Paris: La haute Science, 1953.

Anawati, Georges C. *Drogues et médicaments dans l'Antiquité et le Moyen Âge*. Cairo: Dar-al-Maaref, 1959.

Artelt, Walter. "Die ältesten Nachrichten über die Sektion menschlicher Leichen im mittelalterlichen Abendland." In *Abhandlungen zur Geschichte der Medizin und der Naturwissenschaften*, Heft 34. Berlin: Schadewalt, 1940.

——. "Arzt und Leibesübungen in Mittelalter und Renaissance." *Medizinhistorisches Journal*, 3 (1968), 222-242.

Aschner, Bernard. "Empiricism and Rationalism in Past and Present Medicine." *Bulletin of the History of Medicine*, 17 (1945), 269-286.

A condensed but comprehensive survey of the dilemma since Antiquity, with some anti-rationalist bias.

Astruc, Jean. *Mémoires pour servir à l'histoire de la Faculté de Médecine de Montpellier*. Paris: Cavelier, 1767.

Baader, Gerhard. "Zur Anatomie in Paris im 13. und 14. Jahrhundert." *Medizinhistorisches Journal*, 3 (1968), 40-53.

An examination of Ricardus Anglicus' *Anatomia* and the pseudo-Galenic *Anatomia vivorum*, which leads the author to the conclusion that the scholastic method was hindering the progress of anatomy at Paris.

——. "Zur Überlieferung der lateinischen medizinischen Literatur des frühen Mittelalters." *Forschung, Praxis, Fortbildung*, 17 (1966), 139-141.

Barbera, Juan Saiz. *Raimundo Lulio; Genio de la Filosofia y Mystica Española*. Madrid: E.P.E.S.A., 1963.

Bariéty, Maurice and Charles Coury. *Histoire de la Médecine*. Paris: Arthème Fayard, 1963.

Bashford, Henry. "Gilbert and John." In *The Harley Street Calendar*, pp. 1-22. London, 1929.

On Gilbertus Anglicus and John of Gaddesden.

Batllori, Miguel. "Arnau de Vilanova antiscolastique d'après les textes catalans et italiens." *Bibliotheca Pontificii Athenaei Antoniani*, 7 (1951), 567-581.

——. "Arnau de Vilanova en Italie, 1267?-1276?" *Analecta Sacra Tarraconensia*, 23 (1950), 83-101.

——. "Nuevos datos biográficos sobre Arnaldo de Vilanova." *Archivo Iberoamericano de Historia de la Medicina*, 8 (1956), 235-237.

——. "Orientaciones bibliograficas para el estudio de Arnau de Vilanova." *Pensamiento*, 10 (1954), 311-323.

——. "La documentación de Marsella sobre Arnau de Vilanova y Joan Blasi." *Analecta Sacra Tarraconensia*, 21 (1948), 75-119.

Baumel, Jean. *Histoire d'une seigneurie du Midi de la France*. 2 vols. Montpellier: Causse, 1969-1972.

This richly documented study covers the period from 985 to 1349.

Beaujouan, Guy. "Manuscrits médicaux du moyen âge conservés en Espagne." *Mélanges Casa Velázquez*, 8 (1972), 161-221.

Beccaria, Augusto. *I codici di medicina del periodo presalernitano*. Rome: Istituto di Storia della Medicina, 1956.

Beek, Henri. *Waanzin in de Middeleeuwen; Beeld van de Gestoorde en Bemoeienis met de Zieke*. Haarlem, Neth.: De Toorts, 1969.

According to a note of the publisher, in 1969 Dr. Charles H. Talbot planned an English translation of this remarkable and copiously documented Leiden thesis whose author died less than one month before receiving his M.D. degree. The translation has not yet appeared in print.

Beusing, H. H. *Leben und Werke des Richardus Anglicus*. Leipzig dissertation, 1922.

Birkenmajer, A. "Le rôle joué par les médecins et les naturalistes dans la réception d'Aristote aux XIIe et XIIIe siècles." In *La Pologne au VIe Congrès International des Sciences Historiques*, pp. 123-134. Warsaw, 1930.

Bloch, Marc. *The Royal Touch; Sacred Monarchy and Scrofula in England and France*. Tr. by J. E. Anderson. London: Routledge & Kegan Paul, 1973.

Botto Micca, A. "Le relazione scientifiche tra l'Università di Montpellier ed i medici piemontesi." *Minerva Medica*, 56 (1965), 848-853.

Brody, Saul N. *The Disease of the Soul; Leprosy in Medieval Literature*. Ithaca, N.Y.: Cornell University Press, 1974.

Brunel, Clovis. "Recettes médicales du XIIIe siècle en langue de Provence." *Romania*, 7 (1962), 3-10.

Brunhölz, Franz. "*De more medicorum*; Ein parodistisch-satirisches Gedicht des 13. Jahrhunderts." *Sudhoffs Archiv*, 39 (1955), 289-315.

Brunn, Walter von. "Die Stellung des Guy de Chauliac in der Chirurgie des Mittelalters." *Archiv für Geschichte der Medizin*, 14 (1921), 65-106.

——. "Die Bedeutung Salernos für die Medizin." *Neue Jahrbücher für das klassische Altertum*, 45-46 (1920), 381-394.

Bullough, Vern L. *The Development of Medicine as a Profession; The Contribution of the Medieval University to Modern Medicine*. Basel: Karger, 1966.

——. "The Development of the Medical Guilds at Paris." *Mediaevalia et Humanistica*, 12 (1958), 33-40.

——. "The Development of the Medical University at Montpellier to the End of the Fourteenth Century." *Bulletin of the History of Medicine*, 30 (1956), 508-523.

——. "The Medieval Medical University at Paris." *Bulletin of the History of Medicine*, 31 (1957), 197-211.

——. "Status and Medieval Medicine." *Journal of Health and Human Behavior*, 2 (1961), 12-23.

——. "The Teaching of Surgery at the University of Montpellier in the Thirteenth Century." *Journal of the History of Medicine and Allied Sciences*, 15 (1960), 202-203.

——. "Training of the Nonuniversity-Educated Medical Practitioners in the Later Middle Ages." *Journal of the History of Medicine and Allied Sciences*, 14 (1959), 446-458.

Campbell, Anna M. *The Black Death and Men of Learning*. New York: Columbia University Press, 1931.

Campbell, Donald. *Arabian Medicine and Its Influence on the Middle Ages*. 2 vols. London: Kegan Paul, 1926.

Seems to overstate Arabic influences.

Carreras y Artau, Joaquin. "Arnaldo de Vilanova, apologista antijudaico." *Sefarad*, 7 (1947), 49-61.

——. "La llibreria d'Arnau de Vilanova." *Analecta Sacra Tarraconensia*, 9 (1933), 63-84.

——. *Relaciones de Arnau de Vilanova con los Reyes de la casa de Aragon.* Barcelona: Real Academia de Buenas Letras, 1955.

Castiglioni, Arturo. *Storia della medicina.* 2 vols. Verona: Mondadori, 1948.

Unlike too many textbooks of medical history, this work reflects an appreciation of the medieval contribution and it also has an extensive bibliography.

Catalogue des manuscrits de la Bibliothèque de la Ville et de celle de l'École de Médecine de Montpellier. Paris: Imprimerie nationale, 1849.

Chevalier, A. "Constantinus Africanus and the Influence of the Arabs on Salerno." *Ciba Symposia*, 5 (1944), 1725-1731.

—— and John Gerlitt. "Famous Medical Teachers of Montpellier." *Ciba Symposia*, 2 (1940), 412-417.

Cholmeley, Henry P. *John of Gaddesden and the Rosa Medicinae.* Oxford: Clarendon, 1912.

Chrohns, Hjalmar. "Zur Geschichte der Liebe als Krankheit." *Archiv für Kulturgeschichte*, 3 (1905), 66-86.

Clements, R. D. "Avicenna, the Prince of Physicians." *Minnesota Medicine*, 48 (1965), 1567-1568; 49 (1966), 187-192.

Colomer, Eusebio. "Las *Artes Liberales* en la concepción cientifica y pedagogica de Ramón Llull." In *Arts libéraux et philosophie au Moyen Âge; Actes du IVe Congrès International de Philosophie Médiévale*, pp. 683-690. Montreal, 1969.

Combes, J. "Quelques remarques sur les bourgeois de Montpellier au Moyen Âge." In *Mélanges Pierre Tisset*, ed. by Société d'histoire du droit, pp. 93-132. Montpellier: Faculté de droit et des sciences économiques, 1970.

Cooper, Sonoma. "The Medical School of Montpellier in the Fourteenth Century." *Annals of Medical History*, 2 (1930), 164-195.

Curtius, Ernst R. *European Literature and the Latin Middle Ages.* Tr. by Willard R. Trask. New York: Harper & Row, 1963.

De Bra, Felix. "Kurzer Abriss der Geschichte der medizinischen Schule von Montpellier; Ein historischer Versuch." *Janus*, 10 (1905), 297-306, 352-363, 406-418, and 416-472.

This essay has value for its emphasis on historiography, but it is still based on the assumption that obscurantism stifled all inquiry during the Middle Ages.

Delaruelle, Étienne; M.-H. Vicaire; et al. *Les Universités du Languedoc au XIIIe siècle.* Cahiers de Fanjeaux no. 5. Toulouse: Privat, 1970.

A collection of stimulating colloquium papers by historians of the Midi who deal with the development of the Languedoc *studia generalia* primarily in the context of religious history.

Delhaye, Philippe. "La place des arts libéraux dans les programmes scolaires du XIIIe siècle." In *Arts libéraux et philosophie au Moyen Âge; Actes du IVe Congrès International de Philosophie Médiévale*, pp. 161-173. Montreal, 1969.

Delmas, Paul. *Diplômes de l'ancienne Université de Médecine de Montpellier.* Geneva: Droz, 1925.

While concentrating on diplomas of 1605, this brief paper contains an interesting introduction on graduating procedures since the thirteenth century.

——. *La Faculté de Médecine de Montpellier; Guide du visiteur.* Montpellier: Faculté de Médecine, 1926.

This guide is useful because it relates the present buildings of the Faculty to its earlier history.

—— and Charles Guérin-Valmale. "L'hippocratisme montpelliérain; ses origines; Rôle prépondérant des Fils de l'Islam et des Enfants d'Israel." *Aesculape*, 13 (1923), 2-6, 36-41, 53-57, 90-95, 110-113, 135-138, and 163-165.

A detailed study which emphasizes the Greco-Latin as well as the Judeo-Arabic influences, but overlooks the "native" element in Montpellier's tradition. Also, the authors often jump to seemingly unwarranted conclusions and supply no bibliographical references.

Demaitre, Luke. "The Idea of Childhood and Child Care in Medical Writings of the Middle Ages." *Journal of Psychohistory*, 4 (1977), 461-490.

——. "Nature and the Art of Medicine in the Later Middle Ages." *Mediaevalia*, 2 (1976), 23-47.

——. "Scholasticism in Compendia of Practical Medicine, 1250-1450." *Manuscripts*, 20 (1976), 81-95.

——. "Theory and Practice in Medical Education at the University of Montpellier in the Thirteenth and Fourteenth Centuries." *Journal of the History of Medicine and Allied Sciences*, 30 (1975), 103-123.

Denifle, Heinrich. *Die Entstehung der Universitäten des Mittelalters bis 1400.* 2 vols. Berlin: Weidmann, 1885.

Diels, H. "Über die Schrift Antipocras." *Sitzungsberichte der preussischen Akademie der Wissenschaften*, 22 (1916), 376-394.

Diepgen, Paul. "Arnald von Villanova und die Medizin des Mittelalters." In *Lychnos; Lärdomshistoriska Samfundets Årsbok*, pp. 222-242. Uppsala and Stockholm, 1939.

——. *Arnald von Villanova als Politiker und Laientheologe.* Berlin and Leipzig, 1909.

——. "Die Bedeutung des Mittelalters für den Fortschritt in der Medizin." In *Essays on the History of Medicine, presented to Karl Sudhoff*, pp. 99-120. Oxford and Zurich: Seldwyla, 1924.

——. *Frau und Frauenheilkunde in der Kultur des Mittelalters.* Stuttgart: Thieme, 1963.

——. "Der Kirchenlehrer Augustin und die Anatomie im Mittelalter." *Centaurus*, 1 (1951), 206-211.

——. "Studien zu Arnald von Villanova." *Archiv für Geschichte der Medizin*, 3 (1909), 115-130, 188-196; 4 (1910), 369-396; 5 (1911), 88-120; and 6 (1913), 380-391.

——. "Studien zu Arnald von Villanova." In *Medizin und Kultur*, pp. 108-185. Stuttgart, 1938.

——. *Traum und Traumbedeutung als medizinisch-naturwissenschaftliches Problem im Mittelalter*. Berlin: Springer, 1912.

——. *Über den Einfluß der autoritativen Theologie auf die Medizin des Mittelalters*. Wiesbaden: Steiner, 1958.

Dilleman, Georges. "La pharmacopée au Moyen Âge; I. Les ouvrages." *Revue d'Histoire de la Pharmacie*, 19 (1968), 199-211.

Dock, George. *Printed Editions of the Rosa Anglica of John of Gaddesden*. Reprint from *Janus*, 12 (1907).

Doutrepont, Georges. "La littérature et les médecins en France." *Bulletin de l'Académie royale de Langue et de Littérature françaises de Belgique*, 12 (1933), 137-177.

Dulieu, Louis. "Arnaud de Villeneuve et la médecine de son temps." *Montpellier médical* (3e série), 63 (1963), no. 1.

——. *La Chirurgie à Montpellier de ses origines au début du XIXe siècle*. Avignon: Presses universelles, 1975.

Contains little on the Middle Ages beyond a list of surgeons.

——. "La diététique et la nutrition à Montpellier à travers les âges." *Monspeliensis Hippocrates*, 46 (1969), 4-16.

——. "La pharmacie à Montpellier à travers les âges. *Monspeliensis Hippocrates*, 3 (1960), 3-11.

——. "Réflexions sur les origines de l'École de médecine de Montpellier." *Histoire de la Médecine*, numéro spécial (1959), 35-55.

——. "L'Université de Médecine de Montpellier et l'Église." *Monspeliensis Hippocrates*, 5 (1962), 3-9.

Durling, Richard J. "An Early Manual for the Medical Student and the Newly-fledged Practitioner: Martin Stainpeis' Liber de modo legendi seu studendi in medicina ([Vienna] 1520)." *Clio medica*, 5 (1970), 7-33.

Ebied, R. Y. *Bibliography of Mediaeval Arabic and Jewish Medicine and Allied Sciences*. London: Wellcome Institute of the History of Medicine, 1971.

Elsässer, Günther. "Ausfall des Coitus als Krankheitsursache in der Medizin des Mittelalters." In *Abhandlungen zur Geschichte der Medizin und der Naturwissenschaften*, Heft 3. Leipzig, 1934.

Emery, Richard W. "The Black Death of 1348 in Perpignan." *Speculum*, 42 (1967), 611-623.

Euzière, Jules. "Les Maîtres de l'École de Médecine de Montpellier et la philosophie." *Monspeliensis Hippocrates*, 23 (1964), 13-23.

A general survey of seven centuries, which concludes with a sweeping encomium of the "personalism" of Montpellier's tradition.

Forgue, Émile. *Sept siècles de chirurgie à la Faculté de Médecine de Montpellier*. Montpellier: Faculté de Médecine, 1920.

Fort, George F. *Medical Economy during the Middle Ages; A Contribution to the History of European Morals from the Time of the Roman Empire to the Close of the 14th Century*. New York: Kelley, 1970.

This is a reprint of the 1883 edition, with a new introduction by Dr. Morris H. Saffron.

Gabrieli, F. "La medicina arabe e la Scuola di Salerno." *Salerno*, 1.3 (1967), 12-23.

García Ballester, Luis. *Historia social de la medicina en la España de los siglos XIII al XVI*. Vol. 1. *La minoria musulmana y morisca*. Madrid: Akal, 1976.

Ganszyniec, R. *Brata Mikolaja z Polski Pisma Lekarskie* [The Medical Writings of Brother Nicholas of Poland]. Poznan, 1920.

Garrison, Fielding H. *An Introduction to the History of Medicine*. With Medical Chronology, Suggestions for Study, and Bibliographic Data. 4th ed. Philadelphia: Saunders, 1960.

While convenient and commonly used as a textbook in courses on medical history (where still taught in schools of medicine), this work, first edited in 1929, is often unreliable and needs to be revised.

Germain, Alexandre. *Histoire de la Commune de Montpellier*. 3 vols. Montpellier: Martel, 1851.

Goichon, A. M. *The Philosophy of Avicenna and Its Influence on Medieval Europe*. Tr. with notes and a preface by M. S. Khan. Delhi: Motilal Banarsidass, 1969.

Goltz, Dietlinde. "Versuch einer Grenzziehung zwischen 'Chemie' und 'Alchemie'." *Sudhoffs Archiv*, 52 (1968), 30-47.

Gordon, Benjamin Lee. *Medieval and Renaissance Medicine*. New York: Philosophical Library, 1959.

The sources used for this popularized historical survey are rather dated: they include Edward Gibbon, G. C. Coulton, Donald Campbell, and so on.

Graham, Thomas F. *Medieval Minds; Mental Health in the Middle Ages*. London: Allen & Unwin, 1967.

Grmek, Mirko D. "Influence of Avicenna on Western Medicine in the Middle Ages." *Salerno*, 1 (1967), no. 4: 7-21.

Guillemin, Bernard. "Les Français du Midi à la Cour pontificale d'Avignon." *Annales du Midi*, 74 (1962), 29-38.

Guitard, Eugène-Humbert. "Les attaques de Guiot de Provins contre les praticiens instruits à Montpellier au XIIe siècle." In *XVIe Congrès international d'Histoire de la Médecine; Résumé des communications*, p. 78. Montpellier, 1958.

Gurlt, Ernst J. *Geschichte der Chirurgie und ihrer Ausübung; Volkschirurgie, Alterthum, Mittelalter, Renaissance*. 3 vols. Berlin: Hirschwald, 1898.

Though dated, this manual is still useful because of its lucid organization and its summaries of medical writings.

Hain, Ludwig. *Repertorium bibliographicum saeculi XV*. 4 vols. Milan, 1948 (reprint of the 1826-1838 edition).

Halevy, Mayer A. "Contribution à la protohistoire de la médecine montpelliéraine; Ibn Zabarra et les relations hispano-languedociennes au XIIIe siècle." In *XVIe Congrès international d'Histoire de la Médecine; Résumé des communications*, pp. 15-16. Montpellier, 1958.

Hall, Thomas S. *Ideas of Life and Matter*. Chicago: University of Chicago Press, 1969.

——. "Life, Death and the Radical Moisture; A Study of Thematic Pattern in Medieval Medical Theory." *Clio Medica*, 6 (1971), 3-23.

Hamarneh, Sami. *Bibliography on Medicine and Pharmacy in Medieval Islam*. Stuttgart: Wissenschaftliche Verlagsgesellschaft, 1964.

——. *Origins of Pharmacy and Therapy in the Near East*. Tokyo: Naito Foundation, 1973.

Handerson, Henry E. *Gilbertus Anglicus; Medicine of the Thirteenth Century, with a Biography of the Author*. Cleveland: Cleveland Medical Library Association, 1918.

Hariz, J. *La part de la médecine arabe dans l'évolution de la médecine française*. Paris dissertation: Faculté de Médecine, 1922.

Harrison, R. K. "Healing Herbs of the Bible." *Janus*, 50 (1961), 9-54.

Hartmann, F. *Die Literatur von fruh- und hoch-Salerno*. Leipzig: Binschwanger, 1919.

Hauréau, Bernard. "Arnaud de Villeneuve, Médecin et chimiste." In *Histoire littéraire de la France*, 28 (1881), 26-126 and 487-490.

Heidel, William A. *Hippocratic Medicine, Its Spirit and Method*. New York: Columbia University Press, 1941.

Henslow, G. *Medical Works of the Fourteenth Century, together with a List of Plants Recorded in Contemporary Writings, with their Identification*. London: Chapman & Hall, 1899.

Useful primarily because of the list of plants.

Hillgarth, J. N. *Ramon Lull and Lullism in Fourteenth-Century France*. New York: Oxford University Press, 1972.

Hirsch, August, et al., edds. *Biographisches Lexikon der hervorragenden Ärzte aller Zeiten und Völker*. 6 vols. Berlin and Vienna, 1929-1936.

Hoffmann, Gerda. "Beiträge zur Lehre von der durch Zauber verursachten Krankheit und ihre Behandlung in der Medizin des Mittelalters." *Janus*, 37 (1933), 129-144, 179-192, and 211-217.

Holcomb, R. C. "The Antiquity of Congenital Syphilis." *Bulletin of the History of Medicine*, 10 (1941), 148-167.

Huard, Pierre, and Mirko D. Grmek. *Mille ans de chirurgie en Occident: v^e-xve siècles*. Paris: Roger Dacosta, 1966.

The high quality illustrations make this work uniquely valuable.

Illgen, Herbert O. *Die abendländischen Rhazes-Kommentatoren des XIV. bis XVII. Jahrhundert* ("Auszug aus der Inaugural-Dissertation"). Leipzig: Lehmann, 1921.

Irsay, Stephen d'. "The Black Death and the Mediaeval Universities." *Annals of Medical History*, 7 (1925), 220-225.

——. *Histoire des Universités françaises et étrangères des origines à nos jours*. 2 vols. Paris: Picard, 1933-1935.

——. "The Life and Works of Gilles de Corbeil." *Annals of Medical History*, 7 (1925), 362-378.

——. "Les sciences de la nature et les universités médiévales." *Archeion, Archivio de la Storia della Scienza*, 15 (1933), 216-231.

——. "Teachers and Textbooks of Medicine in the Medieval University of Paris." *Annals of Medical History*, 8 (1926), 234-239.

A valuable attempt to supply information about the curriculum between 1274 and 1395, a period for which no official book lists are extant for Paris.

Jarcho, Saul. "Galen's Six Non-Naturals: A Bibliographic Note and Translation." *Bulletin of the History of Medicine*, 44 (1970), 372-377.

The author concludes that the list of six factors influencing health was derived from Galen, but that the term 'non-naturals' was probably coined later.

Jarry, Daniel. "Diététique et hygiène aux XIIIe et XIVe siècles." *Languedoc Médical*, 41 (1958), 5-24.

Jaulmes, Paul and Rose. "La rectification de l'alcocl, invention languedocienne." *Monspeliensis Hippocrates*, 46 (1969), 17-29.

Joly, Robert. "La traduction latine du livre II du Régime et l'unité de la tradition hippocratique." *L'Antiquité classique*, 33 (1964), 45-57.

From a critical comparison of texts, namely a Latin translation of the 6th century and the two oldest extant Greek MSS, the author concludes that the Latin translations deviated from the original text earlier than has been thought.

Kibre, Pearl. "The Faculty of Medicine at Paris, Charlatanism, and Unlicensed Medical Practices in the Later Middle Ages." *Bulletin of the History of Medicine*, 27 (1953), 1-20.

——. "Hippocrates Latinus: Repertorium of Hippocratic Writings in the Latin Middle Ages." *Traditio*, 31 (1975), 32 (1976), 33 (1977), 34 (1978), 35 (1979), to be continued.

——. "Hippocratic Writings in the Middle Ages." *Bulletin of the History of Medicine*, 18 (1945), 371-412.

——. *The Nations in the Mediaeval Universities*. Cambridge, Mass.: Mediaeval Academy of America, 1948, repr. 1962.

——. "The *Quadrivium* in the Thirteenth Century Universities (with Special Reference to Paris)." In *Arts libéraux et philosophie au Moyen Âge; Actes du IVe Congrès international de Philosophie médiévale*, pp. 175-192. Montreal, 1969.

——. *Scholarly Privileges in the Middle Ages; The Rights, Privileges, and Immunities, of Scholars and Universities at Bologna, Padua, Paris, and Oxford.* Cambridge, Mass.: Mediaeval Academy of America, 1962.

Kirsch, Eberhard. "Avicennas Lehren von den Sexualleiden." In *Melemata; Festschrift für Werner Leibbrand*, pp. 49-56. Mannheim, 1967.

Klebs, A. C. "Incunabula scientifica et medica; Short Title List." *Osiris*, 4 (1938), 1-89.

Koch, M. P., and G. Kiel. "Die spätmittelalterliche Gesundheitslehre des 'Herrn Arnolds von Mumpelier'." *Sudhoffs Archiv*, 50 (1966), 361-374.

Kottek, Samuel. "L'art de conserver la santé. Essai de comparaison entre un texte de Maimonide (*Hilchot Deot*, chap. 4) et le *Regimen Sanitatis* de Salerne." *Revue d'Histoire de la Médecine hébraïque*, 87 (1970), 5-9.

Kristeller, Paul O. "Beiträge der Schule von Salerno zur Entwicklung der scholastischen Wissenschaft im 12. Jahrhundert." *Artes Liberales; Studien und Texte zur Geistesgeschichte des Mittelalters*, 5 (1959), 84-90.

——. *Latin Manuscript Books before 1600*. 3rd edition. New York: Fordham University Press, 1965.

——. "Nuove fonti per la medicina Salernitana del secolo XII." *Rassegna Storica Salernitana*, 18 (1958), nos. 1-4.

——. "The School of Salerno; Its Development and Its Contribution to the History of Learning." *Bulletin of the History of Medicine*, 17 (1945), 138-194.

Lain Entralgo, Pedro. *Mind and Body; Psychosomatic Pathology: A Short History of the Evolution of Medical Thought*. Tr. by Aurelio M. Espinosa. New York: Kenedy, 1956.

Lawn, Brian. *The Salernitan Questions; An Introduction to the History of Medieval and Renaissance Problem Literature*. Oxford: Clarendon, 1963.

Leibbrand, Werner. *Heilkunde; eine Problemgeschichte der Medizin*. Freiburg: K. Alber, 1953.

Surveys the philosophical background of medical thought in Greece, at Salerno, in the school of Chartres, as well as the question of the universals, Scholasticism, and so on. It is documented with numerous excerpts from the sources.

—— et al. *Kompendium der Medizingeshichte*. Munich: Banaschewski, 1964.

Lennox, William G. "John of Gaddesden on Epilepsy." *Annals of Medical History*, 1 (1939), 283-307.

Levey, M. "Medical Deontology in Ninth-Century Islam." *Journal of the History of Medicine and Allied Sciences*, 21 (1966), 358-373.

Littré, Émile. "Henri de Mondeville, un des chirurgiens du Roi Philippe le Bel." In *Histoire littéraire de la France*, 28 (1881), 325-352.

Llinarès, Armand. "Les conceptions physiques de Raymond Lulle: de la théorie des quatre éléments à la condamnation de l'alchimie." *Études philosophiques*, 22 (1967), 439-444.

——. *Raymond Lulle; Philosophe de l'action*. Paris: Presses universitaires de France, 1953.

Louis, A. "Analyse des bibliothèques de deux médecins malinois du XVe siècle." *Janus*, 53 (1966), 241-305.

MacKinney, Loren C. "Dynamidia in Medieval Medical Literature." *Isis*, 24 (1936), 400-414.

——. *Early Medieval Medicine: with Special Reference to France and Chartres*. Baltimore: Johns Hopkins University Press, 1937.

——. "Medical Education in the Middle Ages." *Journal of World History*, 2 (1955), 833-855.

——. "Medical Ethics and Etiquette in the Early Middle Ages: The Persistence of Hippocratic Ideals." *Bulletin of the History of Medicine*, 26 (1952), 1-31.

——. *Medical Illustrations in Medieval Manuscripts*. London: Wellcome Institute for the History of Medicine, 1965.

——. "Medieval Medical Dictionaries and Glossaries." In *Medieval and Historiographical Essays in Honor of J. W. Thompson*, ed. by J. L. Cate and E. N. Anderson, pp. 240-268. Chicago, 1938.

Major, Ralph H. *A History of Medicine*. 2 vols. Springfield, Ill.: Charles Thomas, 1954.

Makdisi, George. "The Scholastic Method in Medieval Education; An Inquiry into Its Origins in Law and Theology." *Speculum*, 49 (1974), 640-661.

Martin, Henriette. "À propos d'un précieux manuscrit de Pétrarque de la Faculté de Médecine de Montpellier." *Monspeliensis Hippocrates*, 3 (1960), 3-7.

McVaugh, Michael R. "'Apud Antiquos' and Medieval Pharmacology." *Medizinhistorisches Journal*, 1 (1966), 16-23.

——. "Arnald de Villanova and Bradwardine's Law." *Isis*, 58 (1967), 56-64.

——. "An Early Discussion of Medicinal Degrees at Montpellier by Henry of Winchester." *Bulletin of the History of Medicine*, 49 (1975), 57-71.

——. "The Humidum Radicale in Thirteenth-Century Medicine." *Traditio*, 30 (1974), 259-283.

——. *The Mediaeval Theory of Compound Medicines*. Unpublished Ph.D. dissertation, Princeton University, 1964.

——. "Quantified Medical Theory and Practice at Fourteenth-Century Montpellier." *Bulletin of the History of Medicine*, 43 (1969), 397-413.

——. "Theriac at Montpellier, 1285-1325 (with an edition of the *Questiones de tyriaca* of William of Brescia." *Sudhoffs Archiv*, 56 (1972), 113-144.

——. "Two Montpellier Recipe Collections." *Manuscripta*, 20 (1976), 175-180.

Mendelsohn, E. *Heat and Life*. Cambridge, Mass.: Harvard University Press, 1964.

Meyer, Paul. "Traités en vers provençaux sur l'astrologie et la géomancie." *Romania*, 26 (1897), 225-275.

Meyer-Steineg, Theodor, and Karl Sudhoff. *Illustrierte Geschichte der Medizin*. 5th edition by Robert Herrlinger and Fridolf Kudlien. Stuttgart: Fischer, 1965.

Michler, Markwart. "Guy de Chauliac als Anatom." *Medizinische Monatschrift*, 18 (1964), 306-311.

Miller, Joseph M., et al. *Readings in Medieval Rhetoric*. Bloomington, Ind.: Indiana University Press, 1973.

Møller-Christensen, V. "Evidence of Tuberculosis, Leprosy and Syphilis in Antiquity and the Middle Ages." In *xixth International Congress of the History of Medicine* [Basel, 1964]; *Aktuelle Probleme*, pp. 229-237. Basel: Karger, 1966.

Montoliu y de Togores, Manuel de. *Ramon Lull i Arnau de Vilanova*. Barcelona: Alpha, 1958.

Mor, Carl Guido. "À l'origine de l'École de Montpellier: Rogerius ou Placentin?" *Recueil de mémoires et travaux de la Société d'histoire du droit écrit*, 6 (1967), 11-22.

Morton, Leslie T. *A Medical Bibliography (Garrison and Morton): An Annotated Check-List of Texts Illustrating the History of Medicine.* 3rd ed. Philadelphia: Lippincott, 1970.

Muntner, Suessman. "Un traducteur hébreu du Aboul-Kasim traitant de l'instruction et de l'éthique des médecins de la France méridionale au XIIIe siècle." In *XVIe Congrès international d'Histoire de la Médecine; Résumé des communications*, p. 21. Montpellier, 1958.

A note on Shemtov ben Jizchaq or Schem Tob ben Isaac (1196–ca. 1265).

Neuburger, Max. *History of Medicine.* Tr. by Ernest Playfair. 2 vols. London: Oxford University Press, 1925.

While this textbook perpetuates several misconceptions about medieval medicine, its last pages contain a useful survey of vernacular medical literature.

——. "Vis Medicatrix Naturae." *Medical Life*, 39 (1932), 657-692.

Niebyl, Peter H. "The Non-Naturals." *Bulletin of the History of Medicine*, 45 (1971), 468-492.

Traces the development of the idea of "non-natural" medical elements from Galen through the 19th century.

Niedling, Justus. *Die mittelalterlichen und frühzeitigen Kommentare zur Techne des Galenos.* Leipzig dissertation, 1924.

Ober, W. B., and A. J. Conway. "John XXI, Ophthalmologist, Professor of Physic, Pope." *New England Journal of Medicine*, 273 (1965), 39-40.

Ogden, Margaret S. "The Galenic Works Cited in Guy de Chauliac's *Chirurgia magna*." *Journal of the History of Medicine and Allied Sciences*, 28 (1973), 24-33.

——. "Guy de Chauliac's Theory of the Humours." *Journal of the History of Medicine and Allied Sciences*, 24 (1969), 272-291.

Oliver, F. "Relación entre los médicos aragonenses, catalanes y navarros con la célebre Universidad de Montpellier." *Anales de Medicina y Cirurgia*, 46 (1966), 223-228.

O'Malley, Charles D., ed. *The History of Medical Education.* Berkeley: University of California Press, 1970.

Paetow, Louis J. *The Arts Course at Medieval Universities, with Special Reference to Grammar and Rhetoric.* Urbana, Ill.: University of Illinois Press, 1910.

Pagel, Julius. *Geschichte der Medizin.* 2 vols. Berlin: Karger, 1898.

Though dated, this is still one of the most balanced surveys.

Pagès, P. "Zyklen und Rhythmen im Hippokratismus von Montpellier." *Medizinische Welt*, 11 (1965), 558-564, 612-614, 654-657, 706-713.

Although dealing primarily with the 19th century, this study illustrates the remarkable vigor of Montpellier's Hippocratic tradition.

Paniagua, Juan A. "L'Arabisme à Montpellier dans l'œuvre d'Arnau de Vilanova." *Scalpel*, 117 (1964), 631-637.

——. "Arnau de Vilanova, medico escolastico." *Asclepio*, 18-19 (1966-1967), 533-552.

——. "La médecine arabe à Montpellier dans l'œuvre d'Arnau de Vilanova." In *XVI*[e] *Congrès international d'Histoire de la Médecine; Résumé des communications*, p. 23. Montpellier, 1958.

——. *Estudios y notas sobre Arnau de Vilanova*. Madrid: Consejo superior de investigaciones scientificas, 1963.

——. "Importancia europea de la medicina de Arnau de Vilanova." *Archivo Iberoamericano de Historia de la Medicina*, 8 (1956), 305-314.

——. "Vida de Arnaldo de Vilanova." *Archivo Iberoamericano de Historia de la Medicina*, 3 (1951), 3-83.

Pansier, Pierre. "Catalogue des MSS médicaux des bibliothèques de France. I. MSS latins des bibliothèques autres que la Bibliothèque Nationale de Paris." *Archiv für Geschichte der Medizin*, 2 (1908-1909), 1-46 and 385-432.

——. "Documents pour servir à l'histoire de la Faculté de Médecine de Montpellier au Moyen Âge." *Monspeliensis Hippocrates*, 2 (1959), 5-12.

This is a reprint of an article in *Montpellier médical*, 22 (1905), which summarizes, with minor corrections, the following study.

——. "Les maîtres de la Faculté de Médecine de Montpellier au Moyen Âge." *Janus*, 9 (1904), 443-451, 499-511, 537-545, 593-602; and 10 (1905), 1-11, 58-68, 114-125.

This systematic list of Montpellier masters loses most of its value because the author indiscriminately included persons whose connections with the faculty of medicine cannot be documented.

——. "Jean de Tournemire; Étude bio-bibliographique." *Monspeliensis Hippocrates*, 42 (1968), 5-12 (reprint from *Mémoires de l'Académie de Vaucluse*, [2e série] 4 [1904], 89-102).

——. "La pratique de l'ophthalmologie dans le Moyen Âge latin." *Janus*, 9 (1904), 3-26.

Parenti, F. "Analisi dei rapporti fra medicina, magia e superstizione nel medioevo." *Pagine di storia della Medicina*, 10 (1966), 47-51.

Payne, Joseph F. "Arnold de Villa Nova on the Therapeutic Use of Human Blood." *Janus*, 8 (1903), 432-435 and 477-483.

Perlitz, Walter. *Die Augenheilkunde des Arnold von Villanova*. Berlin: Schade, 1904.

Pignol, Jean. "Géraud du Berry et l'École de Médecine de Montpellier au commencement du 13e siècle." *Annales du Midi*, 11 (1889), 395-397.

This Gerardus may have been associated with Paris rather than with Montpellier.

Pinès, Jacques. "Des médecins juifs au service de la Papauté du XIIe au XVIIe siècle." *Revue d'histoire de la Médecine hébraïque*, 59 (1965), 123-132.

Pinto, Lucille B. "The Folk Practice of Gynecology and Obstetrics in the Middle Ages." *Bulletin of the History of Medicine*, 47 (1973), 513-522.

Pirkner, E. H. "Epilepsy in the Light of History." *Annals of Medical History*, 1 (1929), 453-480.

Portigliotti, Giuseppe. "Dissections clandestines au XIVe siècle." *Aesculape*, 13 (1923), 284-286.

Examining the trial of one master Albertus who in 1319 exhumed a corpse, the author concludes that there were no dissections before those by Mundino at Bologna (1306-1310) and that, even thereafter, there were many objections to dissection. However, the author fails to emphasize the fact that, at least in the case of master Albertus, the exhumation rather than the dissection was the crime.

Poulle, Emmanuel. "Un instrument astronomique dans l'occident latin: la 'saphea'." In *A Giuseppe Ermini*, 1: 491-510. *Studi medievali*, no. 10.1-3. Spoleto: Centro Italiano di Studi sull'alto Medioevo, 1970.

Puschmann, Theodor. *A History of Medical Education from the Most Remote to the Most Recent Times*. Tr. and ed. by Evan H. Hare. New York: Hafner, 1966 (facsimile of the 1891 edition).

Ranking, G. S. A. "The Life and Works of Rhazes." In *XIIth International Congress of Medicine*, pp. 237-268. London, 1913.

Rashdall, Hastings. *Universities of Europe in the Middle Ages*. Rev. ed. by F. M. Powicke and E. B. Emden. 3 vols. London: Oxford University Press, 1936.

Rath, Gernot. "Gilles de Corbeil as Critic of his Age." *Bulletin of the History of Medicine*, 38 (1964), 133-138.

——. "Montpellier en face du Moyen Âge allemand." In *XVIe Congrès international d'Histoire de la Médecine; Résumé des communications*, p. 25. Montpellier, 1958.

——. "Montpellier im Urteil des deutschen Mittelalter." In *Fachliteratur des Mittelalters; Festschrift für Gerhard Eis*, pp. 307-310. Stuttgart: Metzler, 1968.

Rather, L. J. "The 'Six Things Non-Natural': A Note on the Origins and Fate of a Doctrine and Phrase." *Clio Medica*, 3 (1968), 337-347.

Renan, Ernest. "Armengaud, Fils de Blaise, Médecin." In *Histoire littéraire de la France*, 28 (1881), 127-138.

Rennau, Therese. *Die Gynäkologie des Arnald von Villanova mit Erläuterungen*. Freiburg-im-Breisgau dissertation, 1912.

Riddle, John M. "The Introduction and Use of Eastern Drugs in the Early Middle Ages." *Sudhoffs Archiv*, 49 (1965), 185-198.

——. "Lithotherapy in the Middle Ages: Lapidaries considered as Medical Texts." *Pharmacy in History*, 12 (1970), 39-50.

——. "Pomum ambrae; Amber and Ambergris in Plague Remedies." *Sudhoffs Archiv*, 48 (1964), 111-122.

——. "Theory and Practice in Medieval Medicine." *Viator*, 5 (1974), 157-184.

Robbins, Rossell Hope. "Medical Manuscripts in Middle English." *Speculum* 45 (1970), 393-415.

——. "The Physician's Authorities." In *Studies in Language and Literature in Honour of Margaret Schlauch*, pp. 335-341. Warsaw, 1966.

Robinson, Victor. "Rhazes." *Medical Life*, 28 (1921), 206-218.

Roos, Heinrich. "Le Trivium à l'université au XIIIe siècle." In *Arts libéraux et philosophie au Moyen Âge; Actes du IVe Congrès international de Philosophie médiévale*, pp. 193-197. Montreal, 1969.

Rosen, Edward. "Did Roger Bacon invent Eyeglasses?" *Archives internationales d'histoire des sciences*, 26 (1954), 3-15.

——. "The Invention of Eyeglasses." *Journal of the History of Medicine and Allied Sciences*, 11 (1956), 13-46 and 183-218.

Rosenthal, F. "'Life is short, the Art is long': Arabic commentaries on the First Hippocratic Aphorism." *Bulletin of the History of Medicine*, 40 (1966), 226-245.

——. "An Eleventh-Century List of the Works of Hippocrates." *Journal of the History of Medicine and Allied Sciences*, 28 (1973), 156-165.

Roussel, P. "Montpellier, rivale de Salerne." *Galeno*, 15 (1967), 6-9.

Rubin, Stanley. *Medieval English Medicine*. New York: Barnes and Noble, 1974.

The focus of this work is on early medieval England, and the chapters on the 13th and 14th centuries are quite brief and general.

Russell, Josiah Cox. "L'évolution démographique de Montpellier au Moyen Âge (tr. by Philippe Wolff)." *Annales du Midi*, 74 (1962), 345-360.

Sa'di, Lutfi M. "A Bio-Bibliographical Study of Hunain ibn Ishaq al-Ibadi (Johannitius, 809-877 A.D.)." *Bulletin of the History of Medicine*, 2 (1934), 409-446.

Sarton, George. *Introduction to the History of Science*. 3 vols. in 5. Washington & Baltimore: Williams & Wilkins, 1929-1948.

Sayous, André, and Combes, J. "Les commerçants et les capitalistes de Montpellier au XIIIe et au XIVe siècles." *Revue historique*, 153 (1940), 33-44.

Schipperges, Heinrich. *Die Assimilation der arabischen Medizin durch das lateinische Mittelalter*. Wiesbaden: Steiner, 1964.

——. "Die frühen Übersetzer der arabischen Medizin in chronologischer Sicht." *Sudhoffs Archiv*, 39 (1955), 53-93.

——. "Medizinischer Unterricht im Mittelalter." *Deutsche medizinische Wochenschrift*, 85 (1960), 856-861.

——. "Melancolia als ein mittelalterlicher Sammelbegriff für Wahnvorstellungen." *Studia generalia* 20 (1967), 723-736.

——. "Eine noch nicht veröffentlichte 'Summa medicinalis' des Petrus Hispanus in der Biblioteca Nacional zu Madrid." *Sudhoffs Archiv*, 51 (1967), 187-189.

Schneider, Wolfgang. *Lexikon alchemistisch-pharmazeutischer Symbole*. Weinheim: Verlag Chemie, 1962.

Although it is far from exhaustive, this dictionary of alchemical symbols and terms is both useful and fascinating.

Segondy, Jean. "Le Pape Urbain V, insigne bienfaiteur de l'Université de Montpellier." *Monspeliensis Hippocrates*, 23 (1964), 7-13.

——. "Saint Roch de Montpellier." *Monspeliensis Hippocrates*, 23 (1964), 3-7.

Seidler, Eduard. *Die Heilkunde des ausgehenden Mittelalters in Paris: Studien zur Struktur der spätscholastischen Medizin*. Wiesbaden: Steiner, 1967.

———. "Die Spätscholastik im Urteil der Medizingeschichte." *Archiv für Geschichte der Medizin*, 48 (1964), 299-322.

Shatzmiller, Joseph. "Notes sur les médecins juifs en Provence au Moyen Âge." *Revue des Études juives*, 28 (1969), 259-266.

Shaw, James R. "Scientific Empiricism in the Middle Ages: Albertus Magnus on Sexual Anatomy and Physiology." *Clio Medica*, 10 (1975), 53-64.

Siegel, R. E. "Clinical Observation in Hippocrates; An Essay on the Evolution of the Diagnostic Art." *Journal of the Mount Sinai Hospital of New York*, 31 (1964), 285-303.

Sigerist, Henry E. "Bedside Manners in the Middle Ages; The Treatise De cautelis medicorum, attributed to Arnold of Villanova." *Quarterly Bulletin of Northwestern University Medical School*, 10 (1946), 138-143.

———. "Early Mediaeval Medical Texts in Manuscripts of Montpellier." *Bulletin of the History of Medicine*, 10 (1941), 27-47.

———. "Masse und Gewichte in den medizinischen Texten des frühen Mittelalters." *Kyklos*, 3 (1930), 439-444.

Simon, Isidore. "Les médecins juifs et la fondation de l'École de Médecine de Montpellier." *Revue d'histoire de la médecine hébraïque*, 68 (1965), 79-93.

Singer, Charles. "Sur un manuscrit attribué à Guy de Chauliac." In *Comptes rendus du IIe Congrès international d'Histoire de la Médecine*, pp. 168-176. Évreux, 1922.

The text, edited and discussed here by Singer, is actually an excerpt, garbled but verbatim, from Bernard de Gordon's *Liber pronosticorum*.

Siraisi, Nancy G. *Arts and Sciences at Padua: The Studium of Padua before 1350*. Toronto: Pontifical Institute of Mediaeval Studies, 1973.

———. "The *Expositio Problematum Aristotelis* of Peter of Abano." *Isis*, 61 (1970), 321-339.

———. "The *Libri Morales* in the Faculty of Arts and Medicine at Bologna: Bartolomeo da Varignana and the Pseudo-Aristotelian Economics." *Manuscripta*, 20 (1976), 105-118.

———. "The Music of Pulse in the Writings of Italian Academic Physicians (14th and 15th centuries)." *Speculum*, 50 (1975), 689-710.

Soubiran, A. "L'Islam conservatoire de la médecine." *Aesculape*, 48 (1965), 2-45.

Stannard, Jerry. "The Herbal as a Medical Document." *Bulletin of the History of Medicine*, 43 (1969), 212-220.

Sticker, G. "Lepra im Mittelalter." In *XVIIth International Congress of Medicine*, pp. 37-39. London, 1913.

Steinschneider, Moritz. "Constantinus Africanus und seine arabischen Quellen." *Archiv für pathologische Anatomie und Physiologie*, 37 (1866), 351-410.

This study, while groundbreaking, was largely superseded by Lynn Thorndike's discussion of Constantinus in *A History of Magic and Experimental Science*, 1 (New York, 1923), 742-759.

———. *Die hebräischen Übersetzungen des Mittelalters und die Juden als Dolmetscher*. Graz: Akademische Druck- und Verlagsanstalt, 1956 (reprint of the 1893 edition).

Storch, E. P. von, and T. J. C. von Storch. "Arnold of Villanova on Epilepsy." *Annals of Medical History*, 10 (1938), 251-260.

Sudhoff, Karl. "Constantin, der erste Vermittler muslimischer Wissenschaft ins Abendland und die beiden Salernitaner Frühscholastiker Maurus und Urso als Exponenten dieser Vermittlung." *Archeion*, 14 (1932), 359-369.

——. *Erstlinge der pädiatrischen Literatur*. Munich: Münchner Drucke, 1925.

——. "Fünf ansteckende Krankheiten." *Mitteilungen zur Geschichte der Medizin und der Naturwissenschaften*, 14 (1915), 405-406.

——. "Gedanken über die Ausbildung chirurgischer Operateure an den Hochschulen im 13. Jahrhundert." *Mitteilungen zur Geschichte der Medizin und der Naturwissenschaften*, 17 (1918), 294-295.

——. "Genetische Zusammenhänge und regionale Bedingtheiten in der Medizin des 12. und 13. Jahrhunderts." *Archiv für Geschichte der Medizin*, 21 (1929), 63-74.

——. "Zur Geschichte der Lehre von der kritischen Tagen im Krankheitsverlaufe." *Archiv für Geschichte der Medizin*, 21 (1929), 1-22.

——. "Medizinischer Unterricht und seine Lehrbehelfe im frühen Mittelalter." *Archiv für Geschichte der Medizin*, 21 (1929), 28-37.

——. "Pestschriften aus den ersten 150 Jahren nach der Epidemie des 'schwarzen Todes' 1348." *Archiv für Geschichte der Medizin*, 4 (1910), 191-222, 389-424; and 5 (1911), 36-87.

——. "Zum *Regimen sanitatis Salernitanum*." *Archiv für Geschichte der Medizin*, 8 (1915), 344-355; 9 (1916), 202-222; 10 (1917), 88-96; 11 (1918), 111-123; and 12 (1919), 155-167.

——. "Salerno, eine mittelalterliche Heil- und Lehrstelle am tyrrhenischen Meere." *Archiv für Geschichte der Medizin*, 21 (1929), 43-62.

——. "Salerno, Montpellier und Paris um 1200; Ein Handschriftenfund." *Archiv für Geschichte der Medizin*, 20 (1928), 51-62.

Examines the concluding verses of a commentary on Gilles de Corbeil's *De urinis*, whose last chapter contained the famous invective against Montpellier.

——. "Die Verse Isidors von Sevilla auf dem Schrank der medizinischen Werke seiner Bibliothek." *Archiv für Geschichte der Medizin*, 21 (1929), 23-27.

Szpilczynski, Stanislaw. "Considérations sur les conceptions pseudoscientifiques des pratiques de Nicolas de Pologne (XIIIe siècle)." In *XVIe Congrès international d'Histoire de la Médecine; Résumé des communications*, pp. 30-31. Montpellier, 1958.

Tabanelli, Mario. *Un Secolo d'oro della chirurgia francese (1300)*. Vol. 1. *Henri de Mondeville*; vol. 2. *Guy de Chauliac*. Publications of *Collana di Monografia di 'Romagna Medica'*, nos. 26 and 28. Forli: Valbonese, 1969-1970.

Talbot, Charles H. *Medicine in Medieval England*. London: Oldbourne, 1967.

Teissier, Léon. "La cité de Montpellier au temps de Raymond Lulle." *Revue des Pays d'Oc*, 2 (1933), 384-390.

Temkin, Owsei. *The Falling Sickness; A History of Epilepsy from the Greeks to the*

Beginning of Modern Neurology. 2nd ed., rev. Baltimore: Johns Hopkins University Press, 1971.

——. *Galenism; Rise and Decline of a Medical Philosophy*. Ithaca, N.Y.: Cornell University Press, 1973.

——. "Geschichte des Hippokratismus im ausgehenden Altertum." *Kyklos*, 4 (1932), 14-34.

Thorndike, Lynn. "The Date of the Translation by Ermengaud Blasius of the Work on the Quadrant by Profatius Judaeus." *Isis*, 26 (1937), 306-309.

——. *A History of Magic and Experimental Science*. 8 vols. New York: Columbia University Press, 1923-1958 (reprint 1960-1964).

——. "Notes on Medical Texts in Manuscripts at London and Oxford." *Janus*, 48 (1959), 141-202.

——. "The Relation between Byzantine and Western Science and Pseudo-Science before 1350." *Janus*, 51 (1964), 1-48.

——. "Rufinus: A Forgotten Botanist of the Thirteenth Century." *Isis*, 18 (1932), 63-76.

——. "Some Alchemical Manuscripts at Bologna and Florence." *Ambix*, 5 (1956), 107.

——. "Three Texts on Degrees of Medicines (*De gradibus*)." *Bulletin of the History of Medicine*, 38 (1964), 533-537.

——. "The Three Latin Translations of the Pseudo-Hippocratic Tract on Astrological Medicine." *Janus*, 49 (1960), 103-129.

——. "Traditional Medieval Texts concerning Engraved Astrological Images." In *Mélanges Auguste Pelzer*, pp. 260-267. Louvain: Nauwelaerts, 1947.

——. "Translations of Works of Galen from the Greek by Nicoló da Reggio (c. 1308-1345)." *Byzantina Metabyzantina*, 1 (1946), 213-235.

——. "Translations of Works of Galen from the Greek by Peter of Abano." *Isis*, 33 (1942), 649-653.

——. "Vatican Latin Manuscripts in the History of Science and Medicine." *Isis*, 13 (1929), 53-102.

——, and Pearl Kibre. *A Catalogue of Incipits of Mediaeval Scientific Writings in Latin*. Augmented and rev. ed. Cambridge, Mass.: The Mediaeval Academy of America, 1963. Supplemented in two further articles: "Additional Addenda and Corrigenda" in *Speculum*, 40 (1965), 116-122; and "Further Addenda and Corrigenda" in *Speculum*, 43 (1968), 78-114.

Turchini, Jean. "Salerne et Montpellier." *Monspeliensis Hippocrates*, 4 (1961), 3-7.

A panegyric, rather than an historical examination, of the relations between Salerno and Montpellier.

Ullmann, Manfred. *Die Medizin im Islam* (*Handbuch der Orientalistik* ed. by B. Spuler, 1. Abteilung: *Der nahe und mittelere Osten*, Ergänzungsband 6, Abschnitt 1). Leiden and Cologne: E. J. Brill, 1970.

A thorough and scholarly work, to be supplemented by a companion volume by the same author on "Natur- und Geheimwissenschaften im Islam," which I have not yet been able to consult.

Ussery, Huling E. *Chaucer's Physician; Medicine and Literature in Fourteenth-Century England*. Publication of Tulane Studies in English no. 19. New Orleans: Tulane University, 1971.

Vaccaro, Leopold. "The School of Salerno." *Medical Life*, 22 (1929), 271-284.

Vandenbussche, L., with P. Braeckman and S. Top. *Ziekten en Remedies in de Volksgeneeskunde*. Menen, Belgium: Vandenbussche, 1973.

Valuable as an illustration of the continuity of several medieval therapeutic practices.

Verrier, René. *Études sur Arnaud de Villeneuve, 1240(?)-1311*. 2 vols. Leiden: E. J. Brill, 1947-1949.

Wallner, Bjorn. "Drawings of Surgical Instruments in MS Bibl. Nat. Ang. 25 [containing Guy de Chauliac's *Chirurgia Magna*]." *English Studies* (Arngart Anniversary Number), 42 (1965), 182-186.

This brief linguistic note contains a discussion of the nomenclature for the instruments depicted in the Paris illuminated MS.

Walsh, James J. "Clinical Medicine in the Middle Ages: Bartholomew's De proprietatibus rerum." *Medical Life*, 39 (1932), 365-382.

——. *Medieval Medicine*. London: A. & C. Black, 1920.

This popularization is somewhat slight but makes for entertaining background reading, as does the following.

——. *Old Time Makers of Medicine; The Story of the Students and Teachers of the Sciences Related to Medicine during the Middle Ages*. New York: Fordham University Press, 1911.

Walther, Hans. *Proverbia sententiaeque Latinitatis Medii Aevi*. 5 vols., register. Göttingen: Vandenhoeck & Ruprecht, 1963-1967.

Watson, Gilbert. *Theriac and Mithridatum; A Study in Therapeutics*. London: Wellcome Historical Medical Library, 1966.

Wedel, Theodore. *The Medieval Attitude towards Astrology*. Publication of Yale Studies in English no. 60. New Haven: Yale University Press, 1920.

Weisheipl, James. "Classification of the Sciences in Medieval Thought." *Mediaeval Studies*, 27 (1965), 54-90.

——. "The Place of the Liberal Arts in the University Curriculum during the XIVth and XVth Centuries." In *Arts libéraux et philosophie au Moyen Âge; Actes du IVe Congrès international de Philosophie médiévale*, pp. 209-213. Montreal, 1969.

Welborn, Mary Catherine. "The Long Tradition; A Study in Fourteenth-Century Medical Deontology." In *Mediaeval and Historiographical Essays in Honor of J. W. Thompson*, pp. 344-357. Chicago: University of Chicago Press, 1938.

Wenck, Karl. "Johann von Göttingen; Arzt, Bishof und Politiker zur Zeit Kaiser Ludwigs des Bayern." *Archiv für Geschichte der Medizin*, 17 (1925), 141-156.

Wickersheimer, Ernest. "Autour du 'Régime de Salerne'." *Scalpel*, 50 (1952), 1-12.

——. *Dictionnaire biographique des médecins en France au Moyen Âge*. 2 vols. Paris: E. Droz, 1936.

——. "Le discours de réception d'un bachelier en médecine montpelliérain au début du xve siècle." *Bulletin de la Société française d'histoire de la médecine*, 13 (1910), 245-251.

This study of Heinrich Lamme's commencement address shows how Montpellier's medical faculty ca. 1400 gave its own earlier authors, including Bernard de Gordon, a place in the curriculum.

——. *Manuscrits latins de médecine du Haut Moyen Âge dans les bibliothèques de France*. Paris: Centre National de la Recherche Scientifique, 1966.

——. "La question du judéo-arabisme à Montpellier." *Janus*, 21 (1927), 465-473 and, revised and corrected, *Monspeliensis Hippocrates*, 6 (1959), 3-7.

A critical and revisionist survey of the historiography on the Judeo-Arabic contribution to Montpellier's tradition.

——. "Zeitgenossen über den schwarzen Tod." *Archiv für Geschichte der Medizin*, 3 (1910), 348-349.

Withington, Edward T. "Medicine in Mediaeval Monasteries." *Medical Magazine*, 5 (1896), 508-519.

——. "Roger Bacon on the Errors of Physicians." In *Essays on the History of Medicine, Presented to Karl Sudhoff*, pp. 139-157. Oxford and Zurich: Seldwyla, 1924.

Contains the English translation of the *De erroribus medicorum*, written before 1268, in which Bacon criticized the Latin physicians as generally inferior to the Greeks and Arabs.

Wrigley, John E. "A Papal Secret Known to Petrarch." *Speculum*, 39 (1964), 613-634.

This fascinating note on the illness of Pope Clement VI indicates how numerous must have been the staff of physicians at the papal court in Avignon.

Yates, Francis A. "The Art of Ramon Lull." *Journal of the Warburg and Courtauld Institute*, 17 (1954), 115-123.

Zimmerman, J. *Contribution à l'étude des guérisons observées à la suite de la cérémonie du toucher des écrouelles par les rois de France*. Paris dissertation: Faculté de Médecine, 1962.

Though dealing with the same subject as Marc Bloch's *The Royal Touch*, namely scrophula and the kings' healing power, this study is different in approach and focus.

Index of Persons and Places

Subject Index